ORGANIC FERTILIZERS: Which Ones And How To Use Them

ORGANIC FERTILIZERS: Which Ones And How To Use Them

Edited by the Staff of
Organic Gardening and Farming

Rodale Press, Inc.

Book Division
Emmaus, Pa. 18049

Standard Book Number 0-87857-053-5

Library of Congress Card Number 72-93100

PRINTED IN THE U.S.A.
on 100% recycled paper
JB-3

FIRST PRINTING—APRIL 1973

ORGANIC LIVING PAPERBACKS ARE PUBLISHED BY
Rodale Press, Book Division
33 East Minor Street
Emmaus, Pa. 18049

CONTENTS

Chapter

The Organic Fertilizer Success Story

Chapter 1

The fertility of your soil is of vital importance to the growth of your vegetables. And organic matter is of vital importance to your soil's fertility. Nearly all of the nitrogen and sulfur, and more than one-third of the phosphorus that become available for plant use are supplied by organic matter. Smaller quantities of the other plant nutrients also come from this source: consequently, an increase in the rate of organic matter decomposition likewise increases the quantities of nitrogen, phosphorus, potassium, calcium, magnesium and other plant nutrients in the soil solution.

All soils differ widely in their content of plant nutrients. This difference is caused not only by the fact that soils differ greatly in the original content of nutrients which they had but also because soils lose these nutrients through erosion, leaching and the harvesting of crops. Some of these losses are made up by the weathering of minerals, rainfall, action of earthworms and bacteria in the soil and other natural soil phenomena. But serious deficiencies must be corrected if the soil is to produce adequate and healthy crops. This is the reason why organic fertilizers and mineral nutrients should be added—to increase crop yields as well as to produce crops with the proper nutrients.

Gardeners and farmers all over the country are discovering daily that the way to a successful harvest is through the application of organic fertilizers—the replenishing and feeding of soil the way nature intended. From these enthusiastic acclaimers of "the organic way" come grateful tributes and testimonials to *Organic Gardening and Farming* about the powers of such fertilizers. Bountiful yields, greater insect resistance and the defeat of hardpan are just a few of the wonders being wrought by treating the soil with natural nutrients.

The following is a very small sampling of the great number of success stories that constantly come our way:

Hardpan No Longer a Threat

Our soil grew up because of two main ingredients: elbow grease and organic material. I wouldn't swear that one is more important than the other. Liberal dressings of compost or compost material—such as leaves and manure—will visibly change the condition of any soil, but deep cultivation to get the organic matter thoroughly mixed into the soil is essential.

By spading and hoeing mulches and dressings deeply into the earth, we have accomplished two important operations here in California: Addition of the organic matter to feed microscopic animal life, which in turn change the organic matter into nutrients available to the roots. Physically breaking up the soil by spading and cultivating creates millions of tiny air pockets in the soil, giving the microorganisms air to breathe and making the soil fluffy.

Compacted clay soil that dries out to hardpan in the summer is—to an organic gardener, at least—simple precious garden soil with all the organic matter removed. When we moved here, we saw the problem in organic terms. It was just as important to start growing soil as it was to set out our first tomato plant.

The first year we dug up the 10-by-10-foot crabgrass lawn. We had to treat the earth first with long water soaks before we could slice the sod into six- to eight-inch strips. Those slices became the start of my first compost pit at this house. Between the slabs of patio concrete, the excavated garden bed-to-be looked like the bottom of a pan from which a burned chocolate cake had been scraped. The clay bed looked unpromising. Further soakings were necessary to dig the ground. Of course, we were careful not to break down the body of the soil even further by digging it when it was sopping wet. We waited until most of the moisture had drained off and the soil was just about to dry. Turning the clay, we broke it up with a shovel and by breaking clods between our hands.

The barrelful of sand I tried to include would probably have made things worse if I had not added at the same time a pile of unrotted sycamore leaves saved from the previous autumn, a bale of peat moss, a sack of "compost mix" from a factory experimenting with reconstituted garbage, and about eight cubic feet of rotted oak leaves we had collected from the mountains while picnicking. I noticed that the earth's level would rise with each spading. More than giving credit to additions, I believe that the greatest cause of expansion was the spading itself. With each deep spading (I attempted to work down to 18 inches) more air was worked into the composition of the soil.

The first year we grew only tomatoes and cucumbers. That autumn I dug in rotted leaves that my neighbors had thrown away, and I plowed under the dead vines from the summer's crops. I had not become interested in the possibilities of winter gardening at that time, so I allowed the ground to rest and mulched it with leaves raked up during the fall.

Spring came and my spade slipped through the mulch into the ground without too much effort. I felt as if I had a garden for the first time since the year we moved here. There was no sign of the vegetative matter of the summer's crop nor of the first organic applications. Under the mulch, at the line where the leaves began to dissolve into earth, earthworms wriggled.

By the time I added the contents of the compost heap—those crabgrass sods we had lifted a year earlier, plus kitchen refuse, lawn clippings and some leaves rescued from the street cleaners—the soil level had come up about eight inches. We had to start making a terrace of earth at the boundaries of the garden to confine the soil and water when we irrigated.

Since that first year, never once at the beginning or end of a growing season have we failed to add liberal amounts of organic material. And through a system of creating passageways through our densely planted vegetables, we have added organic materials even during the growing season.

My method is to excavate the pathway carefully, throw in the refuse or leaves, and replace the soil over it. Although some few small roots from nearby plants are invariably cut into by this method, I have not found that it noticeably interferes with plant growth. The pathway larded with rotting leaves and refuse becomes a bed for plants during the following season.

As soon as we harvest a crop, the remains of the plants are fed to the compost pile and the mulch is spaded into the soil. In about 30 days, winter or summer, there is little sign of the mulch. Admittedly, however, our California winters are mild.

We mulch winter and autumn crops as well. During the winter months, however, we use only finished compost—whether leaves, garden refuse, or kitchen garbage. Snails, slugs and earwigs seem to love the partially-composted materials, but if we use finished, sweet-smelling compost as a mulch, pests do not seem to be attracted to it.

All the garden is raked and put in trim in the fall and using the resulting refuse along with autumn leaves, we start a new compost pile for the winter. The summer refuse, now fully composted and clean, serves to mulch the pre-winter turnips, potatoes, carrots, beets, leeks, shallots, and lettuces. We find autumn mulching particularly valuable because the winter rains often do not come when one expects them, and those fall droughts can fool the unwary gardener. A mulch helps conserve the water then as it does in summer.

For many years, during midwinter we stopped gardening. Now though, we're never without a winter garden any more because we've found that a lot of vegetables do well despite our light California frosts. But whether we have plants in the ground or not, we find a winter mulch a particularly nice way to get tilth into the soil.

Now if you compare our soil to that of our neighbors around us, it does appear that we have trucked in some rich loam. Our care and feeding of the soil make it fat, and fat soil makes rich produce.

—John J. Meeker

Vigor and Flavor Are the Best Testimonials

It all started with an article in the local newspaper about the local head of the sewerage authority presenting the ladies of a garden club with bags of treated sludge and telling them to go home and put it on their flower beds.

Since much of our garden had already been planted before the arrival of our sludge, we used it to side-dress the rows as the plants became visible. The results became visible after the first rain. Wow!

The bush beans were a dark, healthy green, and withstood the damage from the Mexican beetles much better. The carrots and the red beets have outdone themselves—although thinned once, they may have to be thinned again because they are growing so much. The tomatoes are covered with fruits.

And the cucumbers! The vines are so aggressive they reach out and grab us around the ankles as we walk by. From one 20-foot row we have already picked over 140 slicers as of August 1, and they show no signs of slacking off. What the children don't sell in the neighborhood, I take to the local health food store, where the proprietor has a waiting list of customers wanting organically-grown produce. The corn, pumpkins, cantaloupes, onions and lettuce are more vigorous than any we have grown.

What sludge we didn't apply directly to the garden, was put into the compost box to help reduce the garbage and grass clippings to their proper state. Each time the lawn is mowed, some of the clippings are added to the box. About once a month we go into the box with a tiller, and churn up the whole thing. Each time we have found that the sludge has acted upon whatever else is in the box to break it down.

We have incorporated much organic matter into the garden. Everything from kitchen garbage to the contents of the cat's litter pan and piles contributed by the neighbors' dogs. We have no hardwood trees on our half acre, but a friend contributed 15 large garbage bags of oak leaves this spring. We rented a compost shredder to grind the leaves, and since we were paying $10 a half-day rental, we ground everything in sight, including one garden glove.

Although the sludge is an excellent source of nitrogen, by itself it would not improve soil texture much. Like the clay, it has a tendency to cake when repeatedly wet and then dried. However, when composted with fibrous materials like grass clippings and even a limited amount of pine straw, it is a beautiful sight.

Some people are turned off at the idea of using treated sludge in their gardens. We patiently explain that the plant ingests only the nutrients from the sludge and not the sludge itself. Sometimes we point out that if they knew what went into commercial chemical fertilizers, they would gladly use sludge. We have found that the vigor and flavor of our produce are the best testimonials.

—Ruth W. Miller

Pecans So Rich My Hands Were Oily

Compost you never have enough of it down here in Houston, Texas, where I garden, while leaf mold is also hard to come by, and manure is scarce. I knew I'd have to find a way to compensate for the limited amount of organic materials available hereabouts.

Cottonseed meal did just that. It helped stretch bare handfuls of organic matter to their maximum benefit, and now every plant and tree in my yard receives at least an annual application of this easy-to-work-with plant food and profits from it.

To prepare my vegetable garden for planting, I first work into the soil whatever compost I have, then spread cottonseed meal at the rate of approximately five pounds—about a gallon—per 100 square feet of area and rake it into the top inch or so of the soil. The meal breaks down slowly, releasing nutrients to the plants all during the growing season, and no further feeding is necessary. The only thing I add later is mulch. Yields on everything from parsley to pecans have increased dramatically since I began using cottonseed meal.

Last season I weighed 20 pounds of pecans before and after shelling them. They yielded 60 percent kernels, which were so rich my hands were oily from handling them. The young trees had been fertilized with cottonseed meal each of the two previous winters, in the following way.

A two-inch auger was used to make holes 18 inches deep and apart under the drip lines of the three young trees. Then we sifted 1½ pounds of cottonseed meal for each inch of diameter of the tree (measured about chest height) into the holes, which were then filled with water. While the water was being absorbed I gathered wood ashes, manure, compost and shrimp heads from a fish market and mixed them together in a wheelbarrow. The holes were filled almost to the top with this, then finished with soil—and that was that for another year.

In addition to containing phosphorus, potash and various trace elements, cottonseed meal is rich in nitrogen, which is essential for lush, green growth. While a minimum of 20 leaves is required for a pecan tree to mature a single nut, cottonseed meal encourages the tree to put on lots of leaves, and in turn bear a bountiful crop—if an unusually early hard freeze doesn't come along that year.

Cottonseed meal's low pH makes it ideally suited for use on my acid-loving magnolia and holly trees since the soil here is alkaline. These trees are fertilized in the same way as the pecan trees and at the same time, December or January, when gardening activity is minimal, and before spring growth begins.

Other broad-leaved evergreens, such as azaleas, gardenias and camellias benefit from cottonseed meal also, but they are fed twice a year instead of once, at a different time and in a different way. To feed these shrubs in the winter or early spring would stimulate them to new growth, and they would drop their bloom buds before flowering.

In the spring, I remove all the mulch that hasn't decomposed into the soil, and spread around the base of the plant a handful of cottonseed meal for each foot of height, being careful not to let it come into contact with the trunk. Since the root systems of these shrubs are near the surface, the meal is only scratched in lightly to avoid damaging the roots. Then a new mulch of shredded oak leaves or pine needles is applied, and the plant is given a slow, thorough watering.

About six weeks later I feed them again at the same rate, but this time the meal is sprinkled over the mulch. Doing so prevents the decomposing mulch from robbing the soil of nitrogen, since rainfall or watering will carry nutrients gradually down to the root zone. The last feeding is always given before July 1 so that the new growth will have time to harden before winter.

My lawn also gets its share of cottonseed meal in the spring, but at a time most convenient for me. Timing here isn't crucial. When I can get around to it I use a fertilizer spreader to distribute five pounds of cottonseed meal over each 100 square feet of lawn. Well, that's what I aim for, but it's difficult to be precise in this case. I never bother to water it in because it doesn't burn the grass, and spring rainfall is generally more than adequate here in the Gulf Coast.

This clean, easily applied and easily stored fertilizer does cost more than manure, leaf mold or compost. But where these materials are in short supply, supplementing them with cottonseed meal will increase the yield sufficiently to more than offset the expense. Having to spend money for it is the only disadvantage I've discovered in the several years I've been using cottonseed meal.

—Lois Patterson

Berries Both Plentiful and Flavorful

"A steady diet of bone meal, cottonseed meal and cow manure is the secret of good producing plants," says 72-year-old Ray Reinsmith of Pennsylvania. Reinsmith feeds his plants before and after each growing season, and adds a three-inch mulch of his own compost around the raspberry plants in the late fall after they bear. His compost pile is made up largely of shredded leaves, clippings from the lawn, weeds, stalks, tomato and sunflower plants—all mixed with a generous helping of bone meal and lime.

Reinsmith adds bone meal while transplanting and when setting out new plants. He says the bone meal strengthens the roots, and should be applied at the bottom of the hole before the plant is positioned, and a handful added after the backfill. The plants should be watered heavily for at least a week so they take a firm hold.

His 40 bushes of red raspberries yield up to 80 quarts in the spring, and again as many in the late fall. He has had the same success with 250 plants of strawberries in his half-acre garden which also flourishes with a wide variety of vegetables, flowers and shrubs.

Toward the latter part of November, Reinsmith covers the strawberry plants with two inches of straw after turning the ground over slightly. The plants grow through the straw in the spring, and Reinsmith applies more straw with his mulch and feed in the spring to nourish the plants as well as to keep the weeds down.

The efforts in keeping his plants healthy with the proper year-round diet and protection have made Ray Reinsmith's berries both plentiful and flavorful—and that is certainly worth the extra care.

—Rudy Bednar

A Big Year for Organic Fertilizers

A great transition has happened with natural fertilizers. Slowly but surely, as more and more growers want to use them to build up soil nutrients, they're becoming more widely available. Freight continues to add too greatly to the cost, but local sources of natural soil-building materials are finally bringing down prices by catering to increased regional demand.

Fred Veith has been selling organic fertilizers and soil conditioners for 22 years. He calls his business Nature's Way Products. This year, he has shipped to all 50 states and Canada from his warehouse in Cincinnati, Ohio.

In southeastern Pennsylvania, several firms have been composting materials and blending in rock fertilizers—then distributing these products to sales outlets up and down the East Coast. OGF editor Floyd Allen will soon be reporting on a recent journey that took him to Lithonia, Georgia, where Hybro-tite is mined, and onto Gap, Pennsylvania, where Zook and Ranck mix the potash-rich granite dust with organic materials.

Across the country in Belmont, California, David Pace of the Ecology Trading Center is typical of the many fertilizer distributors (many are listed each month in OGF's "Fertilizer Directory") who hope to reduce prices of materials to organic farmers and gardeners via a quantity or bulk purchasing plan. "We'll deliver a carload of phosphate rock or other material to local farm areas and let the farmers in the area load their trucks directly from the car instead of hauling it miles away from a warehouse or store. Some materials could be purchased for half the usual price or less," Pace says.

"Free fertilizers"—or at least organic materials that are now considered wastes—continue to be a prime source of bringing nutrients to crops. The United States Department of Agriculture has at long last looked at urban wastes as a useful material in crop production. In mid-October, the Agricultural Research Service held a field day in Beltsville, Maryland, specifically to publicize farm applications of sewage sludge. And finally, the President's Water Pollution Control Advisory Board now declares ocean dumping is no longer tolerable and explains how land scarred by strip mining in Shawnee National Forest in southern Illinois benefits from nutrients in sewage sludge.

State experiment stations also stress the value of waste materials to crops: *The effect of cement kiln dust on yield and nutrition of sugar cane.* ("Cement kiln dust, a lime-like byproduct of the cement industry, has been used experimentally by the University of Puerto Rico to increase yields of sugar cane."—Laron E. Golden, Dept. of Agronomy.")

The utilization of cannery wastes. The fruit and vegetable canneries in Santa Clara County, California, produce approximately 125,000 tons of fruit and vegetable wastes a year. Under a program developed by the Cooperative for Environment Improvement (seven canneries in San Jose area), tests have begun on waste applications on a 2,300 acre site—mainly dry-pasture of clay-adobe with a high akaline content. So far, results look good.

Mineral wastes from industry are also being looked at closely. The Nov.-Dec. issue of COMPOST SCIENCE (the Rodale Press bi-monthly journal that specializes in reports of agricultural value of wastes materials) has an article by L. M. Adams, J. P. Capp and D. W. Gillmore on beneficial effects of applying *fly ash* to coal mine spoilbank land that has been stripped bare and left denuded.

But not all fertilizing materials recommended in an organic growing program come out of somebody's waste piles. Many natural fertilizers and soil conditioners have a brand name, are available in everything from a few pound packages to carload lots, and are sold in garden supply stores and farm outlets. You'll probably do best at a place that specializes in natural fertilizers (like the ones listed monthly in OGF's Fertilizer's Directory), so check for the one nearest you—but don't ignore local stores if there's no other alternative.

Here's a quick run-down of sample items, some brands, some general categories: Agrinite (7 percent nitrogen); blood meal; bone meal; chicken manure; Fertilife; Altoona FAM compost; Milorganite; Cold Buster (humates, blended New Mexico "geological compost"); Fertrel; Cottonseed Meal; Dolomite limestone; Greensand Glauconite potash mineral; Hybro-Tite crushed granite; Norwegian seaweed; phosphate rock; fish emulsion; Sea-Born Liquid Seaweed.

One OGF editor gives this report on fertilizer sources within ten miles of our office in Emmaus: "Most people in the area get their commercial types of organic fertilizer from Hofstetter Organic Products in Limeport. Even Agway is sending people over to Hofstetter's farm since its selection is limited to mostly meals and natural rock fertilizers. Sears, Two Guys and other chains have a limited selection of dried manures and mulch materials like wood chips."

From India, where they are vitally concerned with boosting crop yields, comes the report that organic fertilizers are not only much safer than the chemical but that they are "economical by 40 percent." Professor M. V. Bopardikar writes from Poona:

"We in India now have high fertilizer mixtures of 15:15:15 NPK. That is, 45 percent nutrients and 55 percent chlorides, sulphates and so forth, which are detrimental to the soil." Prof. Bopardikar reports that he has been successful replacing these chemicals with a compost base which "in turn reduces the NPK requirements by 20 percent—besides being 20 percent cheaper than the chemical-based fertilizer mixtures. Thus," he concludes, "it is economical by 40 percent."

But you don't have to go to India to practice economy while building soil fertility and upping crop yields. Now that we have to build up our new 300-acre experimental organic farm at Maxatawny in Pennsylvania, we're turning to the municipal composting plant at Altoona which is slightly more than 200 miles away. No matter—we need 2,000 tons of FAM, product of the Altoona plant, and Manager Dan Detwiler has promised to start us off with two 500-ton shipments.

The general swing to natural, non-chemical fertilizers and conditioners on the part of the public has been closely followed by the people who make them. We discovered on questioning them that demand has really gone up, and that it's been a steady increase year after year, an increase that is not regional but one that comes from all parts of the country.

Demand for tree bark, for example, has gone ahead to such an extent that a National Bark Producers' Association has been formed. The first man to tell us about the new organizaion was Paul Esham of Perdue, Inc., which is situated in Maryland, and sells to dealers throughout the East Coast. Perdue specializes in pine bark mulch ranging in size of aggregate from one-half inch to 2½ inches.

Paul Esham's report was confirmed at Greenlife Products, one of whose representatives, Ed Kelley, has been elected president of the newly-formed Bark Association. "Demand has been steady," we were told, "and increasing each year." Greenlife distributes its pine bark mulch from Texas to Maine, and from Minnesota to Florida. Before commencing manufacturing operations in 1962, the company conducted researches concerning pine bark characteristics for eight years. The experiments also included work done at the University of Idaho and experiment stations in Georgia and Florida.

"Now's the time to grow with organics!" is the slogan under which A. H. Hoffman, of Landisville, Pennsylvania, is conducting operations this year. This company also ships west of the Mississippi, up to the Rocky Mountains. "A definite increase in consumer interest and demand" was noted by Milton Schaefer for such diverse organic products as cow manure, cottonseed meal, bone meal, sheep manure, dried blood, peat moss and shell mulch.

The Organic Compost Corporation sells its composted sheep and cow manure on a nation-wide basis to jobbers, distributors and chain stores from its three regional headquarters in Pennsylvania, Wisconsin and Texas. "There has been a steady increase both in production and demand," Marilyn Mueller of OCC reported.

This was confirmed by John Wells of Conrad Fafard, Inc., "Demand is definitely up, and increased production is paying off," he stressed. Fafard is situated in Massachusetts, obtains its peat moss in eastern Canada, its pine bark in the southeastern states, and distributes them east of the Mississippi down to the Gulf Coast.

Associated products such as cocoa bark mulch are also very much in the picture. New distributors have been added by Hershey Estates to handle increased demand for its mulch which, incidentally, is a by-product of its main chocolate staple. The chocolate producers ship their organic bark mulch as far west as Denver from their Pennsylvania headquarters.

"The greatest move in history from the cities to the country and suburbs" was cited by Tom Fitzgerald of Western for "the biggest sustained demand for peat moss in the industry which is coming from all over the country." Western which sells in every state except those on the West Coast, obtains its peat moss from two Canadian sources, Brunswick for the East, and Vancouver for the West.

And so it goes in the organic fertilizer and mulch industry. Demand on the part of an aware public that knew what it wanted has created a market that production is still trying to satisfy. We think that all pioneers in the Organic Movement can take a little credit for this healthy development.

—Jerome Goldstein and Maurice Franz

What You Should Know About Nitrogen

Chapter 2

"I never add nitrogen to my soil," says organic orchardist A. P. Thomson of Virginia. "I have too much, if anything. If you feed the soil organisms that fix nitrogen, they'll feed it to your crops at the slow natural rate they need it."

That sounds like heresy in these days when chemical growers use 400 million tons of factory-made nitrogen fertilizers every year and when even organic growers can get hung up manuring and cover cropping until there's far more nitrogen than their vegetables need. But as environmental scientist Barry Commoner and many others have pointed out, our over-emphasis on this element is causing serious damage to the eco-system as nitrates leach into the water supply, accumulate in plants and animal feed, and cause the ill effects of any nutrient imbalance.

The fact is that a healthy, organic soil builds up stores of nitrates even when left alone. Nitrogen enters the soil in a slow and fairly inefficient manner. Eighty per cent of our atmosphere is nitrogen, so we have lots of nitrogen around. We can only use very little of it, and what we use we call nitrogen-fixing bacteria.

As the nitrogen is brought into the soil there are other microorganisms that make it available as inorganic nitrates in the humus, and the humus becomes the home for the inorganic nitrogen compounds that we need for living matter.

When we have the inorganic nitrate the plants can use it. They take it up from their roots and convert it into proteins. As plants die, the wastes again, the inorganic products, are returned to the soil. The microorganisms go to work, and after a while we have organic nitrates, and the cycle is closed, and on we go.

Or we can build another cycle on it. An animal can come along and eat the plants, and the animal can die and its organic products, living and dead, go back into the soil, and again we keep recycling, natural recycling.

One key about nature's use of nitrogen is that it is very, very slow, and the process of converting from the organic to the inorganic form so plants can use it is done very slowly. Our job, as gardeners, is to help the natural processes which enrich the soil with the right amounts of nitrogen in the right forms.

For example, the woods are healthy, yet no one shoots up the soil with nitrogen-carrying ammonia. And who manures an abandoned field?—yet the grasses and shrubs seem to do fine. The lesson here is that if we start with healthy soil, all we need do is put back on the soil what we take off in crops to maintain that health.

Of course, this doesn't mean putting back the chemicals that scientists have found are contained in plant tissue. It means putting back plant tissue, with its rich mixtures of many elements and compounds, on which soil life feeds and thrives.

And it doesn't mean lumping it on until the plants become saturated with dangerously-high levels of nitrates, the well water becomes polluted, and the rivers run foul. Let's try to get nitrogen into a clear prospective: let's see how much we really need, then where we can get it without damage to the eco-system.

"Cover cropping alone puts back the nitrogen in many cases," writes OGF West Coast Editor Floyd Allen, citing the Japanese and Filipino farmers in his region of California who cover crop with inoculated fava beans, which can add 200 pounds per acre of nitrogen to the soil each year. One successful farmer plants sugar peas for six months on his land, then follows with fava beans for six months. So what the peas remove, the fava beans put back. And this happens even though he cuts and burns the tops of his fava beans. (It would be much better humus-building practice to allow the tops to rot on the ground rather than burning them or even turning them under. A study by William A. Albrecht at the Missouri Agricultural Experiment Station showed that more available nitrogen was returned to the soil if the cover crop was cut and left to decompose rather than turned under.)

A. P. Thomson raises alfalfa and sweet clover on his 37-

acre orchard, containing 1,300 trees, and figures they add 40–50 pounds per acre of nitrogen to his soil, which is plenty for fruit production.

"The whole premise is that a naturally fertile soil supplied with adequate amounts of humus, gives you everything your crops need, including nitrogen. The organic idea is based on fertile, healthy soil, not whether something is raised without chemicals or pesticides, although that is a part of it," Thomson reminds us.

If this is true, that a healthy soil can manufacture enough nitrogen, then why do the chemical farmers add 200–300 pounds of ammonium nitrate per acre?

Read a few ads for chemical nitrogen and you'll see why: "Gets 'Em Off to a Fast Start," reads one. The answer is that overdoses of nitrogen flush the plant with fast growth that looks lush and green and is in fact forced and tasteless. It provides a lot of tissue (and we buy it by the pound) at the expense of quality. Looking again to a woods or old field, the plants here grow slowly, with thrift and hardiness.

Crops in an organic garden get nitrogen from the decomposition of organic matter caused by bacteria and earthworms. In the process, the microbes liberate nitrogen gas into the air. Long ago, when the first life appeared on the earth, there was very little nitrogen in the atmosphere. The fact that the bulk of our atmosphere today is nitrogen is a legacy from eons of working bacteria.

"Managing the nitrate content of the soil means nothing but the application of ripened composts, and growing legume crops in the field rotation," says H. H. Koepf in "Nitrate—An Ailing Organism Calls for Healing."

There's a feeling among chemical agriculturists that composts don't contain enough nitrogen for the garden, that it needs real he-man doses of the element. If they are looking only at the nitrogen content of compost, with its nitrogen still bound up into unusable forms to a large extent, they may be right. But consider compost as food for the bacteria which manufacture nitrates from organic matter, and you have a much truer picture.

Let's look at a typical garden—say it's yours—to see where the nitrogen comes from.

—There are the bacteria in the soil. In a very fertile soil, their weight per acre may be as high as 600 pounds. When they die, their bodies turn to humus, and their proteins are broken down into nitrates usable by plants.

—The azotobacter group of soil microorganisms are free acting, that is, not working in association with the roots of plants. They extract nitrogen from the air. At Cornell University, a study attributed 42 pounds of nitrogen per acre per year to the action of these bacteria.

—Legumes can add from 50 to 200 pounds of nitrogen to the soil per acre per year. In an experiment done at the University of Missouri Agricultural Experiment Station from 1917 to 1941, written up by M. F. Miller, soil patches were kept under various rotation and cover crops to determine nitrogen level variations. A number of researchers worked on the project and they found that sweet clover was the champion nitrogen producer, adding 1,080 pounds of nitrogen per acre over the 24 years. A rotation of corn, wheat and clover held soil nitrogen at steady levels, even adding a little. "All these increases," Miller writes, "with the exception of rye, are substantial ones. A substantial accumulation of organic material should be of benefit to soil productivity even when nitrogen accumulation doesn't keep pace with it." Legumes include all the clovers, alfalfa, cowpeas, fava beans, soybeans, and bean-type plants. One investigator found that legumes, through the nitrogen-fixing bacteria on the root nodules, get two-thirds of their nitrogen requirements from the air. While bacteria can work on chemical fertilizers (or at least on some of them . . . Cyanamid, for instance, inactivates nitrogen-fixing bacteria in the soil), they do so by turning from air-borne nitrogen to the chemicals. But the chemical nitrogens are produced from the air in the first place, using up five tons of coal to produce one ton of nitrogen fertilizer, according to Professor Georg Borgstrom. How much simpler and more ecological to allow the nitrogen-fixing bacteria to work the element from the air. That's why it's good practice to inoculate your legumes with nitrogen-fixing bacteria before planting.

—Earthworms also add nitrogen to the soil, both by converting unusable forms into usable ones and by the nitrogen contained in their dead bodies; dead earthworms may add tons of organic matter to a field every year. Chemical manufacturers and their friends have been assuring the public lately that their products don't kill earthworms. "Experiments have shown that the application of chemical fertilizers actually tends to increase the numbers of bacteria and earthworms," says S. C. Wiggans of the University of Vermont in a recent issue of *Crops and Soils* in an article entitled, "Exploding the Myths of Organic Farming." Wiggans is talking about a 1947 study by Dr. Firman Bear of Rutgers. However, Dr. Bear used a 0–12–12 fertilizer, meaning it contained no nitrogen. Farmer's Bulletin #1569 by the Department of Agriculture says in this regard, "The results of three years application of ammonium sulphate (a common nitrogen fertilizer) have shown incidentally that earthworms were eliminated from the plots where this chemical was used." One teacher of horticulture reacted to the Bear experiment by stating, "My repeated greenhouse tests with earthworms and fertilizers have showed the opposite results."

—Other soil life, including nematodes, mites, millipedes, centipedes, snails, spiders and other insects may total from 200 to 1,000 pounds per acre. Their excreta and dead bodies enrich the soil with nitrogen proteins. Needless to say, insecticides and other chemicals kill off these life forms. Darwin estimated that earthworm and insect excrement amounts to 10 tons per acre in healthy soil. And the Connecticut Experiment Station found that worm castings contain five times as much nitrogen as topsoil.

—Another source of soil nitrogen is the roots of dead plants. In the case of a two-year-old crop of red clover, the roots amounted to three tons per acre and contained 180 pounds of nitrogen. Plants grown in friable, humusy organic soil develop better root systems than on soil wasted by many years of chemicals where aeration is poor.

—The valuable mycorrhiza fungus forms a symbiotic relationship with plant roots. This fungus helps the plant, and in maturity, the roots consume the fungus as a rich source of nutrients. On soils treated with chemicals, the mycorrhiza is absent or reduced in amount.

—Lightning may bring up to 10 pounds of nitrogen to an acre of soil each year, according to one researcher. Westinghouse engineers estimated that more nitrogen is given to the grounds by the world's thunderstorms over a year than is manufactured during that period. Associated with this is rain, which washes down five to six pounds of nitrogen per acre yearly. Disintegrating rocks, if they contain nitrogen, can also add small amounts.

Manure and other organic wastes are rich sources of nitrogen for organic gardeners looking for materials for the compost pile. Composting is the key here, for fresh manures, applied directly to crops, can not only burn them but contain enough soluble nitrogen to cause trouble and leach into local

Dried blood adds nitrogen to stimulate the growth of bacteria but should be used sparingly.

water supplies. The best way to handle manure is to compost it with green matter—about five times as much plant tissue as manure, in layers.

According to Dr. Koepf, "ripened organic fertilizers made with manure, even when applied in high amounts, will not cause toxic nitrate concentrations in plants; rather they will increase the quantity and quality of the crop." Manure will be discussed in further detail in the following chapter.

Other organic nitrogens include:

Tankage, blood meal, dried blood, and fish scrap. They may be had in all parts of the country whether or not you live near the ocean or have a slaughterhouse in your area. Meat tankage consists of waste processed into meal which contains 10 percent nitrogen and up to 3 percent phosphorus. Bone tankage has 10 percent phosphorus and 3 percent nitrogen.

Dried Blood

Blood meal and dried blood contain 15 and 12 percent nitrogen respectively, and 1.3 and 3 percent phosphorus. Blood meal also assays at 0.7 percent potash. These materials may be used directly in or on the planting site, or they may be added to the compost pile. They should be used sparingly because of high nitrogen content—a sprinkling is enough to stimulate bacterial growth. Both are excellent in the compost pile, breaking down green fibrous matter and stimulating general bacterial action.

Bone Meal

Years ago great amounts of buffalo bones were collected on western plains for use as fertilizer; nowadays the main source comes from the slaughterhouse. Consisting mostly of calcium phosphate, the phosphorus and nitrogen content depends mostly on the kind and age of the bone. Raw bone meal has between 2 and 4 percent nitrogen, 22 to 25 percent phosphoric acid. The fatty materials in raw bone meal somewhat delay its breakdown in the soil.

Steamed bone meal contains 1 to 2 percent nitrogen, up to 30 percent phosphorus, available in hardware stores, local nurseries or usually wherever garden products are sold.

Cottonseed Meal

This is made from the cottonseed which has been freed from lints and hulls and then deprived of its oils. (Cottonseed cake is one of the richest protein foods for animal feeding.) Its low pH makes it especially valuable for acid-loving crops. Cottonseed meal analyzes 7 percent nitrogen, 2 to 3 percent phosphorus, and 1.5 percent potash. A truly excellent fertilizer, it is available commercially.

Grass Clippings

Fairly rich in nitrogen, grass clippings are useful when worked into the soil, for adding to compost heaps, or for mulching. Clippings from most lawns contain over one pound of nitrogen and two pounds of potash for every hundred pounds of clippings in the dry state.

Vegetable Residues

If the budget is tight, or you just can't seem to secure animal manures or waste products, you still have the vege-

Vegetable residues are an inexpensive way of supplying nitrogen.

table nitrogens. You can go down to your local brewery and get free dried hops—about 3½ percent nitrogen and 1 percent phosphoric acid. But go in August when the pile is fully dried—its easier handling. Castor pomace at 5½ percent nitrogen is another vegetable by-product that is excellent both in the compost heap and the planting row. Castor pomace is the residue that is left after the oil is extracted from the castor bean. It is handled by fertilizer dealers in various parts of the country.

We use cottonseed meal and soybean meal whenever we turn wood chips or old hay under in the garden. It gives the soil bacteria extra nitrogen to work with. Cottonseed meal offers the extra advantage of a low pH, which makes it fine for acid-loving crops, and is rated at 7 percent nitrogen, 3 percent phosphorus, and 1½ percent potash. It's easily obtained at any grain or feed mill.

Sawdust

A very useful mulch material, sawdust should be used more widely. When plants are about two inches high, a one-inch layer can be applied. Prior to spreading the sawdust, many gardeners side-dress with a nitrogen fertilizer as cottonseed meal, blood meal, tankage, etc. However, many sawdust users do not apply a nitrogen supplement and were satisfied with results. It should be emphasized that this *must* be well rotted sawdust. Raw or pale colored sawdust will mat and cake and prevent proper penetration of rain.

With all these sources of nitrogen for your soil, there's a real danger of overloading it. And that's what has been happening. According to the U.S. Department of Agriculture, about *4 million tons* of chemical nitrogen are applied yearly in this country. Add to that the nitrogen from animal wastes—one scientist found, nitrate concentrations of 5,000 pounds per acre in the subsoil near feedlots—and you can see that in the interest of growing food, we are literally awash in harmful nitrates.

Late last year, the Illinois Pollution Control Board heard testimony—including a statement by OGF Farm Editor Robert Steffen—about the effects of fertilizers with an eye toward determining whether state-imposed restrictions on the rate, timing and application of commercial fertilizers and manures were necessary, the first such hearings in the nation. Also, part of those hearings dealt with the published results of a study by a research team—Barry Commoner, Daniel Kohl and Georgia Shearer—looking into the buildup of nitrates in water supplies and its relationship to the use of nitrogen fertilizers.

Steffen's statement cited research from several sources to back up his claim that chemical fertilizers deplete the fertility of the soil and produce crops that are more susceptible to insect attack and of poorer quality.

The research by Commoner, Kohl and Shearer was conducted in the area of Lake Decatur, the drinking water reservoir for Decatur, Ill. The lake is fed by the Sangamon River watershed which drains a large farming area. Neither the river nor the reservoir receive any significant amounts of nitrogen-carrying sewage from industrial or municipal sources. Their conclusions were that approximately 55 percent of the nitrogen chemicals showing up in Lake Decatur in the spring of 1970 (the period when fertilizers are laid down on crops) comes from fertilizers being washed into the watershed.

The effects of that chemical runoff were discussed in another report, this one on research conducted at Massachusetts Institute of Technology. Four researchers at MIT's department of nutrition and food science said the excessive use of nitrates as fertilizers can lead to high accumulated levels in water supplies and plant tissue. Bacteria can then convert the nitrates into nitrites—the raw material for a group of chemicals known as nitrosamines. Those nitrosamines have been known to cause cancer in animals, and the question is whether animals will take the nitrates and convert them into nitrosamines. In the test tube, at least, the MIT researchers have found that this could happen.

The Commoner research warned that nitrates were causing bodies of water to eutrophy. He mentioned the fact that nitrates from food plants can convert blood hemoglobin to methemoglobin, especially in youngsters, causing death and sickness.

To date, no official action has been taken by the Illinois board.

But the health problems associated with nitrate and nitrite compounds are manifold. It's been pointed out, for instance, that plant proteins formed on soil too rich in nitrogen are less complete than those formed on soil with the right amount.

A study by W. M. Beeson of the Department of Animal Sciences at Purdue in 1964 pointed out that excess nitrates cause both death and toxicity in animals (including man) as well as interfere with the body's ability to convert carotene to vitamin A.

It was William Albrecht who pointed out that chemical farmers "labor under a delusion that a higher concentration of nitrogen in the fertilized vegetation is proof that the grass must contain higher concentrations of proteins . . . but experimental work with rabbits has shown that the body growth of animals varied according to the *fertility* of the soil on which crops for their feed were grown." Rabbits were offered four different hays and corn, and the rabbits consistently chose the hay grown without nitrogen applications.

And farmers who watch their cows closely may notice that they rarely eat the lush, fast-growing grass that forms around their old droppings and which are super-rich in nitrogen. They prefer the grass with lower nitrogen but more vitality.

This brings us full circle, back to the fertility of the soil and the natural amounts of nitrogen that occur in woods and pastures. We admit that composts contain less nitrogen than say, anhydrous ammonia which is over 80 percent pure nitrogen. But in an era when nitrate poisoning of the environment and of people is so advanced that states are beginning to think about stopping heavy nitrogen applications, the organic ideal becomes more and more of a necessity.

Let's sum up by quoting Dr. Koepf:

"There is only one way to keep the plant nutrients available, approximately in the right amounts and pretty much in the right ratio to one another, and that is by the activity of the soil microorganisms; that is, by having sufficient and rich enough organic matter in the soil. It has been recorded that tremendously rich organic soils release several times more soil-born available nitrogen than the plants will use. The forage crops from these soils failed to produce nitrate poisoning in cattle."

Make your compost well, apply it liberally, and rotate legumes through your garden. Those are the simple secrets of nitrogen maintenance in your garden.

—Jeff Cox

Manure: The All-Around Natural Fertilizer

Chapter 3

"I fully share the average farmer's partiality for barn-yard manure in preference to most, if not all, commercial fertilizers."

Sound like a contemporary statement on organic farming? The author is none other than Horace Greeley, legendary newspaperman of the mid-1800's, who had a strong tendency to dabble in agriculture.

"It seems . . . plainly absurd," he went on, "to send [for commercial fertilizer] . . . when this [New York] or any other great city annually poisons its own atmosphere and the adjacent waters with excretions which are of very similar character and value, and which Science and Capital might combine to utilize at less than half the cost. . . ."

And that was written one hundred years ago!

Editor Greeley was fully aware "that a majority of those who would live by its [the soil's] tillage feed it too sparingly and stir it too slightly and grudgingly." He was appalled "that many of our people . . . are squandering money on Commercial Fertilizers . . . Almost any farmer who has cattle, with fit shelter and Winter fodder, can make fertilizers far cheaper than he can buy them."

Greeley insisted that the farmer "cut, when they are in blossom, all the weeds that grow near him, especially by the road-side, cart them at once into his barn-yard, and there convert them into fertilizers . . . pile load after load of freshly-fallen leaves into your yard . . . which the trampling of cattle may readily convert into manure."

Greeley summed up his philosophy of man's relationship to the soil in this way:

"As no deposit in a bank was ever so large that continual drafts would not ultimately exhaust it, so no soil was ever so rich that taking crop after crop from it annually, yet giving nothing back, would not render it sterile or worthless. Sun and rain and wind will do their part in the work of renovation; but all of them together cannot restore to the soil the mineral elements whereof each crop takes a portion, and which, being once completely exhausted, can only be replaced at a heavy cost. Science teaches us to foresee and prevent such exhaustion—in part, by a rotation of crops, and in part by a constant replacement of the minerals annually borne away. . . ."

The farmer "must realize his success depends upon his absolute verity and integrity," warned Greeley. "He deals directly with Nature, which never was and never will be cheated."

—Joanne Moyer

Horace Greeley, however, was not the first, nor the only, one to realize the value of manure as a convenient, cost-free, and ecologically sound fertilizer.

This has been a basic fertilizer used for centuries. Some manures, such as horse, hen, sheep, and rabbit, are considered *hot* manures because of their relatively high nitrogen content. Rabbit manure, for example, analyzes 2.4 N, 1.4 P, and 0.6 K. It's best to allow these manures to compost before applying directly to plants. Cow and hog manure, relatively wet and correspondingly low in nitrogen are called *cold* manures, and ferment slowly. All manures are excellent and should be included in an organic fertilizing program, when available.

Animal manure releases the soil's energy.

Animal manures make things grow in the garden because they stimulate and release a lot of energy in the soil. They are the heart of the compost program, may be fed directly into the growing row without danger, can be used as top-dressing for trees, bush fruits, flower beds, and borders. Whatever you have to fertilize, manure will help you do it better. So it's a good idea to keep rabbits on your place, poultry, goats, or livestock of any kind. If you can't, make every effort to contact a local dairyman, egg farmer, or riding stable. Chances are, the owner will be glad to supply you with this by-product just to get it out of his way.

But be ready to do your own trucking.

It can be obtained from local dairymen, for around $7 a ton—sometimes swap it for the loan of a shredder. One of the two heaps can be used as an earthworm hill to which you can add the family garbage after it is shredded or worked into a slurry. Bacterial and worm activity are so great in those hills that it never really freezes in the winter, they are readily opened to admit more garbage, and then closed over. Like a volcano, it's hot at top but cool down below where the worms stay at a safe distance from the heated part.

Manure has been used by family farmers since man first began to till the soil. But today, many gardeners are turning their backs on it in favor of chemical fertilizers. Ann and George Snopko have not.

Two Or Three Crops Per Season

Over the many years of tending their large family garden, the Snopkos have learned to use cow manure in so many different ways that they find it unnecessary to use any other type of fertilizer. The results they obtain in their garden and orchard are astounding!

Ann and George actually start their garden year in the fall after the harvest is in. All plant refuse is removed from the vegetable patch and flower beds, and put to one side. Fresh manure is then spread evenly over the entire garden plot.

If a soil test shows the soil to be slightly acid, or low in phosphorus or potash, they apply these needed accessory elements before spreading the manure and plowing everything under together. The soil is then left in a rough state until early spring when it is tilled lightly and raked smooth for more effective planting.

After the plowing is finished Ann and George turn their attention to making rich compost for the coming spring and summer. Raked leaves, and any other organic matter they have on hand, are added to the garden refuse; then the combination is layered with fresh cow manure. After the compost is well worked, it is allowed to rest until spring.

George's last task with manure in fall is to feed his semi-dwarf fruit orchard, whcih consists of several different varieties of apple, plum, cherry, apricot, peach and pear trees. He first removes a two-foot circle of turf from around each young tree. Next, he carefully works a reasonable amount of finely crumbled, well-aged cow manure into the exposed topsoil, making certain no manure rests against the bark. This is the only plant food the trees get, yet they produce so heavily that George is often compelled to harvest part of the fruit crop prematurely to prevent limb breakage. The fruits are all of exceptional size and flavor.

Just as their lovely tulips begin to break ground in early spring, the Snopkos resume their faithful use of cow manure. Fresh manure, taken directly from the barn, is placed in a thick layer deep within the hotbed, producing the needed bottom heat for the small vegetable and flower seedlings, already growing in flats on the kitchen windowsill.

Ann, who does most of the spring planting, makes it a practice never to plant *anything* without first adding manure in some form to the soil.

It is during this time that she utilizes the manure-enriched compost that she and George made last fall. When planting seed, such as snap beans or squash, she first drops an inch of compost into each drill or hill, waters it down well, and then drops in the seed which she covers lightly with rich topsoil.

This method of planting—which has long been her secret—gives her almost a 90-percent germination, and some of the strongest, healthiest and best-producing plants I have ever seen. Out of just four 70-foot rows of snap beans (two yellow and two green) she realized *15 bushels* of beans last summer. Some she canned, but the majority she sold to lucky friends and neighbors.

A good handful of this enriched compost also goes into each planting hole when she sets out her tomatoes, cabbage and other crops.

When her spring garden is well on its way, Ann gives all of her plants an extra boost with a liquid plant food made by stirring one-fourth pail of fresh cow manure into three-fourths of a pail of water. This she feeds in varied amounts according to the need for nitrogen. Not only does this extra boost produce bigger-than-average yields, but it also encourages early maturity, enabling Ann to grow a good second crop before frost.

Most of us average one crop of any one vegetable a season. But Ann, through clever planning and with the aid of manure, usually enjoys two or three.

One would think the Snopko garden was taken care of by a dozen highly-skilled gardeners who used the most expensive fertilizers available. But all Ann and George use is that good old-fashioned cow manure that so many gardeners are beginning to turn their backs on. Perhaps we ought to take another look at the many advantages of using manure in our gardens. I, for one, intend to because the Snopkos have convinced me that there is nothing like it.

Horse Manure to the Rescue

About seven years ago, Ralph and Bonnie Kappel of Yellowstone Valley, Montana, were just about ready to give up raising livestock or plants. Then, Manure saved the day. Here's how.

"Our compost pile would be nothing," Bonnie reports, "without horse manure, as it heats up the fastest and best of any manure I've used and I've used chicken, goat, cow and pig manures, too.

"I make about ten tons a year, and use chicken, horse and goat manure as well, mixed with leaves, waste hay and straw. As I grind or shred the stuff, I pile it, wet the layers as I go, leave it three days (it usually heats the first night), turn it, then turn it every other day until it cools off. I tried shredding garbage but found it much easier to feed all the garbage to the goats, chickens or horses, then use their manure in the compost.

"We have used this compost for years now as a mulch on our garden during the summer, then till it in in the fall. We've covered our entire lawn with compost, too, and the amount of water it takes is about half of what it used to be, both on the lawn and garden.

We no longer board our cattle out in the winter, as keeping them here gives us manure to spread on our pasture land. We feed a high quality vitamin-mineral supplement to our livestock, and all our animals are healthy—and so are we. *We no longer spend all our profits on doctor bills for us and our animals either.*

Applies Manure in Fall to Winter Over on Soil

"Horse manure applied in the fall, after all produce is gathered from the garden, seems to help the soil, too. By spring the manure no longer burns the plants. The moisture from the snow works the nitrogen into the soil, and tilling mixes it thoroughly too. We have found that we can pasture our horses in our apple-tree area by hosing in the manure piles every few days.

"It takes considerable water and time, but is better than having bare spots in the grassy areas, or leaving it to draw flies. We use only organic methods of insect control, and the common house fly (as well as the horrible face flies) still gives us and our animals a lot of trouble. We have to keep things as clean as possible during fly season.

"For several years now I have been trying to use all our horse manure by the above-mentioned methods. It may sound silly, but with 50 to 100 bantams, seven goats plus the horses, I just can't use all the manure, even for compost. So another use had to be found last summer as we had a fifth horse here and kept her in a pen near our house. The weather was wet and the smell was terrible. Wet horse manure is impossible to put through a shredder as it only clogs it up.

"A couple years ago some tree choppers dumped huge loads of wood chips and logs in holes left on our land when an old creek bed had been partially filled in with gravel and sand. The chips were rotting very slowly, except where the horses had been wintered in that pasture. There the chips with manure over them were rotting a lot faster than the others.

"We decided to add all our surplus horse manure to those fills of chips. When we irrigate the pasture, the water runs into these low areas, soaking both the chips and manure thoroughly. Because of the gravel underneath there is drainage so the chips don't stay soggy, but just damp and thus act as a compost pile, with much less work on my part. It is really surprising how quickly the grass is moving into these areas, by root runners from the outside edges.

"We have an area of pure sand, left there many years ago by a flood so our next step will be to spread the sand over the chips and manure; then seed it with a good type pasture grass. We'll have to fence the area for a year or so to keep the stock from trampling it or pulling the plants before the roots are well established and the fill has a chance to settle. We're looking forward to a real productive pasture in a couple of years though.

"The ways we have changed have certainly paid big dividends for us in the results that have shown up. I never weed a flower bed anymore, nor do I spade them either; the earthworms do the spading for me; I just apply a new layer of compost to them whenever they need it. I have proved that healthy soil makes healthy plants and healthy plants make healthy animals and humans too. Every one of us should change to the natural way of life."

—Bonnie E. Kappel

Chicken Manure—Good and Cheap

Chicken manure is a high-quality rangeland fertilizer—as good as chemical fertilizers and a lot cheaper. What is more, you're getting rid of the stuff safely and effectively and building the land while you're doing it. Such is the gist of a four-man* report from the University of California

—"Fertilization of Annual Rangeland with Chicken Manure"—which originally appeared in the *Journal of Range Management.*

In a prefatory note, the authors explain that "changing conditions in land use may bring poultry operators into foothill areas and thus provide a cheap source of plant nutrients for rangeland fertilization." While the possibilities appeared excellent—a sort of ecological perpetual-motion-machine was promised—the range operators had some questions they wanted answered before going ahead. These are some of the things about which they wanted definite information:

1—What are the optimum times for application and also the rate of application;

2—How much forage could be expected from manure as contrasted with commercial fertilizer;

3—How long the fertilizer effect would last;

4—How forage quality would be affected;

5—What effect the manure would have on forage legumes;

6—Would fertilization be profitable.

Tests have shown that fertilization with chicken manure increases the protein and phosphorous content of plants.

In the tests that followed in the Santa Ysabel area of San Diego County, California, seven varying treatments were repeated four times on a series of experimental plots: 1—no fertilization; 2—one ton of chicken manure to the acre; 3—two tons; 4—four tons. The report cites the following results:

"Fertilization with either chicken manure or inorganic fertilizer significantly increased first-year forage yields at the study site. In all but one season of application (March 1963) the average response to 1 ton of chicken manure per acre (or $N_{70}P_{40}$) was significantly greater than the average yield for the check plots. On the average, 1 ton of chicken manure resulted in a first-year forage yield increase of 1420 lb., 2 tons of chicken manure increased yields by 2260 lb., and 4 tons of chicken manure resulted in forage yields of 3530 lb. per acre more than the check plots."

The Dollars-and-Cents Aspect

Such returns, the researchers found, permitted the local rangers to show a profit while improving their lands and turning an agricultural glut into a community asset. Here's how the simple arithmetic worked out:

"In rural San Diego County operators of manure spreader trucks will spread poultry manure on land for $3.10 to $4.35 per ton, depending upon distances from poultry farm to areas using the manure.

"A general average of 1600 lb. of extra feed was obtained for each ton of chicken manure. The value of this added feed may be expressed in animal unit months (AUM) or its equivalent (1 AUM = 800 lb. of hay for a 1000 lb. steer). Thus, the 1600 lb. of extra feed is equal to 2 AUM's and has a value of $5.00, using a local average of $2.50 per AUM. (Considerably higher values of an AUM are reported for other annual range areas.) A net profit is, therefore, possible based on local prices of $3.10 to $4.35 per ton (2.5 yards per 1 ton) of poultry manure applied on the range."

So much for the double-entry bookkeeping—what about

the quality of the forage the cheaper stuff was producing. The report states the qualitative case as follows:

"As is the case with range fertilization in general, fertilization with chicken manure produces forage of a higher quality and palatability. Protein and phosphorus content were significantly higher in forage from fertilized plots. Other benefits of fertilization include a longer period of forage availability. Fertilized range is ready to use earlier in the season and, because of its higher palatability as dry feed, may be used with greater efficiency for grazing in the dry summer months.

"The results from this study show that the fertilizer value of chicken manure is equal to equivalent rates of commercial fertilizer. There appears to be a slower release of fertilizer elements from the chicken manure than the inorganic fertilizer but over a 3-year period the net forage response was nearly equal."

—Maurice Franz

Don't Forget the Minerals!

Chapter 4

Nitrogen-rich manures and meals don't tell the whole story. No plant can grow successfully, attain a ripe maturity, or reproduce its own kind without the phosphorus and potash found in the rock minerals. And you can get all you need in local stores at very moderate prices, although the stuff is admittedly heavy and hard to handle.

Although they are considered mainstays of organic gardening and farming, natural rock fertilizers have been more or less neglected by the average gardener. This oversight should be remedied because natural rock fertilizers furnish ample supplies of phosphorus and potash without which *no plant can grow successfully, attain a ripe maturity, or reproduce its own kind.*

Of equal organic significance is the fact that natural mineral fertilizers put these essential nurtients and elements into the soil gradually, literally over the years, so they are absorbed as the plants need them. And because there is no "feast or famine," the soil and plants are in sound balance and the soil structure improves as the nutrients are released.

The basic importance of these elements cannot be overstressed. Their available presence in the soil is absolutely necessary to successful gardening and farming. Phosphorus has been called the "master key to agriculture" because low crop production is due more often to a lack of this element than any other plant nutrient.

Potassium is almost as vital to the well-being of the plant. Root and tuber crops require enormous quantities of potassium, while its sufficient availability insures plump kernels in grains and good straw structure.

What do these facts mean to the average gardener? Just this, he isn't going to grow the kind of organic foods he wants and needs unless he pays strict attention to a sound rock fertilizer program. For real results in the garden and field, we must fertilize the soil with both phosphorus and potassium. Because so much farmland in America is deficient in phosphorus, the farmer is under tremendous pressure to buy mixed chemical fertilizer and superphosphate.

Promotional, informative, and even educational sources repeatedly advise him that he must replace depleted fertility with this wonderful "plant food." He is led to believe that the soil is merely a medium through which you feed the plant with this so-called food. He is told he can grow one corn crop after another on the same field indefinitely, just as long as he supplies the crop with enough chemicals.

Now, the cheapest and best source of phosphorus is finely ground rock phosphate. It should be applied to the soil liberally and as often as it is needed.

Phosphate Rock

Phosphate rock is the commercial term for a rock containing one or more phosphate minerals of sufficient grade and suitable composition for use as fertilizer. Calcium

Rock phosphate is an inexpensive but important source of phosphorous and may be used liberally.

phosphate is the most common mineral carrier. The term includes phosphatized limestones, sandstones, shales and igneous rocks. The rock ranges from 28 to 38 percent phosphoric acid.

Phosphate rock is mined by open-pit methods in all 3 producing fields in the U.S.—Florida, Tennessee and the Western States (Idaho, Montana, Nevada, Utah and Wyoming). Underground mining is limited to the Western States. Colloidal phosphate is a finely divided type of rock phosphate, obtained from the settling ponds used in hydraulic mining of phosphate rock. It usually contains from 18 to 24 percent total phosphoric acid, and its recommended application rate is about 50 percent more than rock phosphate.

Ordinary superphosphate is made by mixing phosphate rock with sulfuric acid. Triple superphosphate is made by mixing phosphate rock with phosphoric acid.

The University of Illinois for years has been the leading advocate of the use of rock phosphate. The farmers in this State still use more tons of rock phosphate than any other State in the Corn Belt.

Here are the first two paragraphs from the Department of Agronomy Sheet Number AG1565I, 3/10/53: "There is some misunderstanding about the availability of phosphorus in phosphate fertilizers. Many farmers have been led to believe that the phosophorus in rock phosphate is largely unavailable to plants and will remain so for many years. That is not the case. This confusion has come from misuse of the ammonium citrate test. The purpose of this test is to check the acid treatment process in making superphosphate, and not to measure the availability of phosphorus in phosphate fertilizers.

"When used on rock phosphate, the ammonium citrate test indicates that the phosphorus in rock phosphate is only very slightly soluble. It is then assumed that only the soluble phosphorus shown by the test is available to plants. *The only real test of availability of phosphorus in a fertilizer is ability of the fertilizer to supply phosphorus to plants.*"

One of the tests run on the Mumford 7 plots at the University was conducted on the Applied-Removed-and-

Remaining basis. 1500 pounds of rock phosphate containing 210 pounds of phosphorus was applied to one plot. 800 pounds of superphosphate containing 70 pounds of phosphorus was applied to another plot. After the plots had been cropped for 15 years, the soil was checked for remaining phosphorus. The rock phosphate plot still had 89 pounds of phosphorus. The superphosphate plot was 42 pounds deficient in phosphorus.

A series of experiments conducted in the Soviet Union reveal that soil fertilized with phosphate rock mixed with fresh manure yields 60 percent more potatoes; 158 bushels per acre, compared to 97. When the manure and phosphate rock were not mixed, but applied separately, 145 bushels were raised. It has been estimated that combining the phosphate and manure increases the availability of the phosphorus to the plant by 150 to 200 percent.

How Much is Available?

Back in 1913, farmers asked Cyril G. Hopkins: "How much of the phosphorus in rock phosphate is available?" Hopkins was one of the really great men in agriculture at the University of Illinois in the old days. He told them *none* is available and if they did not intend to make it available, they should not use it. Truer words were never spoken. A good supply of organic matter in the ground and a constant addition to this supply is the basis of all good farming practices. In the decomposition of organic matter, carbonic acid, nitric acid, and many other organic acids are formed that work upon the mineral particles in the soil, including rock phosphate. This is what makes rock phosphate and all other minerals in the soil available.

Even the most rabid advocate of chemical fertilizer will admit that organic matter is a good thing to have in the soil. In fact it is really needed to make his product function properly. It is a proven fact that to make either the organic method or the chemical method of farming work, we need organic matter. So, if the farmer does a good job of rotating his crops and returns all possible organic matter to the soil, he need have no worry about availability. His rock phosphate will become available naturally, as the crops need it, and it will not become locked up in the soil nor leach away.

70 to 80 percent fine-ground phosphate will pass through a 200-mesh screen having 40,000 holes per square inch. You can't beat it for results—top quality, all-organic crops at the lowest possible cost to the farmer.

But, what about these chemical mixtures now being thrust upon the farmer? Let us examine two such formulae and see what they contain. Take a simple mixed fertilizer, such as 3-12-12, and see just what the farmer is getting. It contains the following:

15 lbs. of 20 percent ammonium sulphate	15 x 20 =	3 lbs.
60 lbs. of 20 percent superphosphate	60 x 20 =	12 lbs.
20 lbs. of 60 percent muriate of potash	20 x 60 =	12 lbs.

95 lbs., plus
 5 lbs. of magnesium limestone or other filler

100 lbs. total mixed chemical fertilizer

A hundred-pound bag contains three pounds of nitrogen, 12 pounds of phosphoric acid, and 12 pounds of potash. Phosphoric acid (P_2O_5) is a compound made up of two parts of phosphorus and five parts of oxygen. Phosphorus makes up 43.66 percent of this compound. Therefore, to reduce phosphoric acid to the *PLANT FOOD ELEMENT PHOSPHORUS*, multiply 12 x 43.66 percent = five pounds of phosphorus.

Potash is also a compound. It is potassium oxide (K_2O), and is made up of two parts of potassium and one part oxygen. Potassium makes up 83 percent of this compound. Therefore, 12 pounds x 83 percent = 10 pounds of potassium.

The nitrogen content is the same, three pounds. Many farmers think the numbers on the bag represent the pounds of so-called plant food. The numbers show what is in the bag, namely, three pounds of nitrogen, 12 pounds of phosphoric acid, and 12 pounds of potash . . . but what the farmer really gets in the plant food element is three pounds of nitrogen, five pounds of phosphorus, and 10 pounds of potassium. The element usually most needed and the one he gets the least of is phosphorus.

Let us break down a hundred pounds of straight superphosphate 0-20-0. The 20 represents 20 pounds of phosphoric acid (P_2O_5). Phosphorus makes up 43.66 percent of this compound. So multiply 20 x 43.66, which gives us 8.7 pounds of phosphorus. This 20 percent superphosphate containing 8.7 pounds of the plant food element phosphorus sells for about $40.00 per ton.

Now we will break down 100 pounds of 32 percent rock phosphate. The 32 represents the pounds of phosphoric acid. So we take 32 x 43.66 percent which gives us 14 pounds of phosphorus. This 32 percent rock phosphate sells for $20.00 per ton. In other words, the total amount of the plant food element phosphorus in one ton of superphosphate is 174 pounds, and this costs the farmer $40.00. The total amount of the plant food element phosphorus in one ton of rock phosphate is 280 pounds, and this costs the farmer $20.00.

The above description of 3-12-12, 0-20-0, and rock phosphate might seem a little repetitious, but we wish to make sure the readers understand what has been stated, and also want to be sure there is no way of having these statements misquoted or misinterpreted. There are many professors and teachers in the agricultural field that do not want to hear anything about a truthful comparison of superphosphate and rock phosphate. When farmers at a farm meeting seek information on these two phosphates, the speakers will go into lengthy orations on the great advantage of using superphosphate because of its availability.

If you're using artificial fertilizers, such as superphosphate or muriate of potash, to build up your soil's mineral content, you're also stuck with the schedule that goes with it—frequent applications of small amounts. But if you're going to use rock phosphate, greensand, granite dust or a mixture of them, then you don't have to think in terms of seasonal or even annual applications to the same fields.

Tests made at the Kentucky Agricultural Experiment Station show that after 26 years, yields of corn, wheat and hay were generally maintained or increased on rock phosphate residual plots. The rock phosphate greatly increased the soil's phosphorus content, and this became gradually available as the rock reacted with various soil components.

The Kentucky tests are only one of many showing the residual quality of natural mineral fertilizers. Here are some results obtained at the Virginia Agricultural Experiment Station several years ago: One ton of rock phosphate spread once in 6 years produced yield increases approximately equal to those produced by 100, 175 and 280 pounds of superphosphate spread every year. Crops used in the test were corn, wheat, and red clover. Corn yields produced by a ton of rock phosphate on a sandy loam soil exceeded those produced by 250 pounds per acre of superphosphate. On an alfalfa plot, one ton of rock phosphate spread once in 6 years was equal to 200 pounds per acre of superphosphate spread annually.

Granite Dust—Big in Potassium

The value of potassium is indicated by its position in the scale of nutritional values; it is the third major plant food. It helps carry carbohydrates throughout the plant system, forms strong stems, and fights disease. It improves the keeping quality of fruits; aids in the production of sugar, starches, and oils; reduces the plants water requirements; is essential to cell division and growth; and helps the plant utilize nitrogen.

Granite dust is an excellent source of organic, slow working potash. Its potash content varies from 3 to 5 percent, but granite dust has been obtained from a Massachusetts quarry with a potash content of 11 percent.

The potash-bearing minerals in granite dust are the potash feldspars and micas, the latter containing the most readily released potash. Small amounts of trace elements are also present.

Commercially processed "fortified" granite dusts are now available, sold as complete natural soil conditioners. Produced on a large scale in the region where they are abundant and packaged, they are then distributed on a nationwide basis.

Just such a product comes from Georgia where the raw granite juts up out of the ground, and 10,000 chicken egg farms are common. The product is a combination of the chicken droppings and pulverized granite, colloidal phosphate, ground meat scraps, and humus composted and heated to 160 degrees, then allowed to cool off gradually. A Pennsylvania-made product combines such organic materials as blood meal, castor pomace, fish meal, cocoa bean shells, phosphate rock, greensand, and seaweeds. It is packaged and marketed in 4 different blends, for maximum effectiveness with different soils and growing problems. These commercial fertilizers or additives may be combined with the gardener's own homemade compost or applied separately.

Two tons of granite dust should be applied to the acre. We believe granite dust should previously be mixed with the manure-phosphate combination, spread, and then turned under. This should be done in the spring, before planting,

Granite dust, an excellent source of potash, is available commercially in different blends for different soils.

when you would normally start working the land. Smaller applications call for 20 pounds to 100 square feet and 200 pounds to 1,000 square feet of this complete fertilizer.

Greensand—Packed with Potash

Greensand is interesting stuff, pleasant to work with. It stays put better than the rock powders, and you can spread some around a tree and go back later to dig it in. Coming as it does from deposits laid down in what was once the ocean, greensand contains not only potash but also trace minerals, silica, iron, and magnesium. It has the property of absorbing and holding water, stimulates helpful soil organisms, and never burns. It is excellent for conditioning both hard and sandy soils, so we saved a bag of greensand to give a special soil conditioning treatment to our grapevines, fruit trees and filbert bushes.

Greensand or greensand mari contains more potash than granite dust; 6 to 7 percent compared with 3 to 5. Being an undersea deposit, it contains most if not all of the elements which are found in the ocean. It is an excellent soil builder.

Superior deposits of greensand contain 50 percent silica, 18 to 23 percent iron oxide, 3 to 7½ percent magnesia, and small amounts of lime and phosphoric acid.

Greensand has the ability to absorb large amounts of water and provides an abundant source of plant-available potash. Its minerals or trace elements are also essential to plant growth. Because it is versatile, it may be applied directly to the plant roots, left on the surface as a combined mulch compost. Once again, combining it with the manure-phosphate rock mixture should be seriously considered, if not put into actual practice.

Once in short supply, greensand is now mined and processed in New York, New Jersey, Maryland and Ohio, whereever the deposits exist, then packaged and distributed over a great part of the country. To these marine sources of soil fertility and texture must be added the new material called humate, which is abundant along the Florida coast and the Gulf of Mexico.

Wood Ashes Also Rich in Potash

Containing 1.5 percent phosphorus, 7 percent or more potash, wood ashes should never be allowed to stand in the rain, as the potash would leach away. They can be mixed with other fertilizing materials, side-dressed around growing plants, or used as a mulch. Apply about 5 to 10 pounds per 100 square feet. Avoid contact between freshly spread ashes and germinating seeds or new plant roots by spreading ashes a few inches from plants. Wood ashes are alkaline.

How to Use Mineral Fertilizers

Rock fertilizers aren't generally hard to ship, handle, store or spread. They can be handled the same way as other bulk fertilizers for farm application. Although they may not always flow like pelleted material, the same spreading machines and other equipment will usually distribute all of them effectively. One important point is that they are not corrosive on metals.

Rock phosphate should always be bought from a source with as low a fluorine content as you can get. Fluorides may become a problem after many years of using any material high in this toxic element.

Generally speaking, the finer the fertilizer is ground, the more available it becomes to the growing plant. This does make the material rather dusty and hard to spread on windy days. If you can get a bulk spreader truck with a hood covering the spinners, it will enable you to do a better job of covering your field, especially on blustery days.

If you receive a bulk carload of rock, it can be stockpiled and spread later. Rain will not cause it to cake. It can be spread with the trailertype spreaders that you might rent. The rock can be spread any time of the year—with autumn usually the best time to incorporate ground-rock fertilizers. The work load is less, the fields are all open, and the heavy traffic on the fields generally causes less compaction because the soil is usually drier. Furthermore, the winter freeze will undo some of the damage from compaction. Rock fertilizers can even be spread on snow, as long as the ground is not frozen and the machines can move where they need to. Fall is similarly the best time of year to spread compost or manure, for pretty much the same reasons.

This preferred timing is really a strong argument in favor of composting—because the manure is stockpiled through the winter and wet spring when one cannot easily get on the fields, and through the summer while growing crops on them. This will give you the compost you need when you should spread it, and in the quantities that you should have.

The discussion above applies pretty much to potash rock as well, or to any other natural sources of mineral fertilizer. Potash rock, however, is generally not as fine as the phosphate and may actually be easier to handle in some respects. What we have seen and used at the farms at Boystown has almost the same characteristics as limestone.

Ground-rock material can, of course, be shipped in closed hoppered cars in bulk, and it is available in sacks, which would enable you to spread it with any grain drill-type spreader. Both the cost and the labor involved would be greater. However, in some cases, this may be the only way you can get it on your land.

One other point that may be brought up here is that if soil tests, tissue tests, and your observations indicate trace mineral deficiencies in your area, this can be a good way of overcoming them. Simply mix your own formula with some rock fertilizer and spread with the grain drill-type spreader. This is more accurate, and the amount of material involved is usually only about 100 pounds per acre of the complete mix, enabling you to put on the very small amounts, and put them where you want them. Generally speaking, this may involve only a pound or two of some material per acre. Mixing these with the more bulky rock is a good way to get your fields covered.

Rock Fertilizer Amounts for the Garden

General recommendations for applying ground rock fertilizer include the following:

Farms—(Per Acre) rock phosphate—1,000 lbs., granite dust—1,500 lbs., greensand—4,000 lbs.

Gardens—(all vegetables) 10 pounds per 100 square feet; or apply directly in hills or rows, with plants and seeds;
Roses and bushes—1 pound each;
Flower beds—5 to 10 pounds per 100 square feet;
Lawns—5 pounds per 100 square feet;
Trees—(all kinds) 15 to 100 pounds per tree. Spread to edge of drip line;
House plants—1 tablespoon to 5-inch pot, 1 teaspoon to smaller pots.

Once a good rock fertilizer broadcasting program is set up, one application will last for three or four years. All the more reason, we now realize, for putting plenty on while you're about it. The stuff is not expensive, and it takes little more time to make a generous application every few years rather than a skimpy yearly spreading. Manures and compost spread several times a year during this period help respark the rock additives and keep them available.

Don't be afraid to use relatively large amounts of natural minerals. They are slow working and cannot burn young plants. If the soil is alkaline, increase the amount of manure either in the mixture or put it directly into the ground before applying the mixture.

Rotating crops and growing legumes, will also reduce an alkaline condition. But if you apply rock phosphate, remember to turn it under thoroughly so the nitrogen-fixing bacteria growing on the legume roots can go to work on it.

Remember that plant symptoms of nutritional deficiency are helpful indicators, but not the whole story. When you observe them, it is too late to correct the situation that year. You can only plan steps for the next to prevent its recurrence.

Mixing fresh animal manures with natural mineral fertilizers may well be a real advance in organic fertilizing. Based on past experience, the various elements work reciprocally, their interactions increasing the potency of each nutrient. Getting this complete fertilizer into the ground promptly reduces nutrient-loss sharply and increases its plant availability.

In all, the use of natural minerals in combination with animal and plant manures is recommended soberly as the cornerstone of a gardening and farming program that will extend over the years. You are putting vital nutrients back into the soil where they will work gradually and naturally to improve soil structure and produce ripened, wholesome, nutritious crops year in and year out.

—Nancy Bubel

ORGANIC SOURCES OF NITROGEN, PHOSPHORUS, AND POTASSIUM
SOME COMMON ORGANIC MATERIALS AND THEIR NPK PERCENTAGES

Material	N	P	K
Activated sludge	5.00	3.00	. . .
Animal tankage	8.00	20.00	. . .
Alfalfa hay	2.45	.50	2.10
Apple leaves	1.00	.15	.35
Bloodmeal	15.00	1.30	.70
Bone meal	4.00	21.00	.20
Brewer's grains (wet)	.90	.50	.05
Castor pomace	5.50	1.50	1.25
Cattle manure (fresh)	.29	.17	.35
Cocoa shell dust	1.04	1.49	2.71
Coffee grounds (dried)	1.99	.36	.67
Corn stalks	.75	.40	.90
Cottonseed	3.15	1.25	1.15
Cottonseed meal	7.00	2.50	1.50
Dried blood	12.00–15.00	3.00	. . .
Duck manure	0.6	1.4	0.5
Fish scrap (red snapper)	7.76	13.00	3.80
Greensand	. . .	1.50	5.00
Ground bone, burned	. . .	34.70	. . .
Hen manure (fresh)	1.63	1.54	.85
Hoof meal and horn dust	12.50	1.75	. . .
Horse manure (fresh)	.44	.17	.35
Immature grass	1.00	.5	1.2
Incinerator ash	.24	5.15	2.33
King crab (dried and ground)	10.00	.26	.06
Lobster shells	4.60	3.52	. . .
Molasses (residue)	.70	. . .	5.32
Oak leaves	.80	.35	.15
Peach leaves	.90	.15	.60
Pear leaves	.7	.12	.4
Pig manure	0.5	0.3	0.5
Rabbit manure	2.4	1.4	0.6
Raspberry leaves	1.35	.27	.63
Red clover	.55	.13	.50
Seaweed	1.68	.75	5.00
Sheep manure (fresh)	.55	.31	.15
Swine manure (fresh)	.60	.41	.13
Tankage	6.00	8.00	. . .
Tobacco stems	2.00	. . .	7.00
Wood ashes	. . .	1.50	7.00

Make Sure Your Soil Has Trace Elements

Chapter 5

Spectacular deficiencies of trace elements occur literally around the world.

In a large part of America, extending from Maine across the top of the country to Washington, iodine deficiencies are widespread, particularly in Montana. There are also extensive areas where the soil is low in manganese, copper and zinc. A lack of boron in plants is found along the Atlantic coastal plain, the northwest Pacific, and also in Wisconsin. Manganese deficiencies are especially acute in Florida, and are also found in the muck soils of Michigan, the Atlantic coastal plain and California. A lack of copper is prevalent in the Great Lakes region, Washington, South Carolina, Florida and California. Recent studies indicate that trace element shortages are more extensive than previously thought, and some are growing even more pronounced because modern agricultural practices withdraw soil reserves without replenishing them.

In areas deficient in trace elements, local flora and fauna have usually made adaptations. Adapted local plants may be healthy, but imported or transplanted species may not do well at all. And those plants which need less of a certain trace element will become dominant as other life forms fail to develop when that element is lacking. One researcher estimates that 50 million acres of croplands require boron fertilization right now and only one-quarter are getting it. In Australia, well-known for its trace-element deficiencies, about 300 million acres of adequately watered land are undeveloped, primarily due to lack of trace elements. These regions, which could be readily reclaimed, would quadruple the present agricultural area of all Australia.

The plant's ability to hold water is affected by trace element nutrition. Radishes receiving fully-balanced mineral nutrition showed a wilting rate of 20 percent in one study, while plants deficient in zinc and copper showed an 80 percent wilting rate under similar conditions.

There are a number of ways to determine whether your soil is deficient in trace elements—but all pose big problems. Most require in-depth scientific analysis, including spectroanalysis, which is expensive. Another way is to visually inspect plants to check for tell-tale signs of trace element shortage. But visual symptoms are of limited value because they only appear when the deficiency is severe, and are of no value in looking for latent deficiencies. Also, simple deficiencies of just one element are rare in nature, and it's more usual to find multiple deficiencies. These alter the visual symptoms considerably when they act in concert so it's very difficult to determine with certainty the cause of an abnormal condition. Rather than looking for symptoms, time would be more wisely spent adding mineral-rich material to the compost pile to make sure the garden has what it needs to grow in health.

The average soil may have more than enough of all the trace elements, and yet years of chemical farming have caused them to be bound into compounds which plants can't use. To get them to go to work, all you need to do is start applying organic matter to the soil.

The most reliable and safe method for releasing micronutrients from the soil, and for adding them in balanced amounts, is by making compost and fertilizing organically with plants that accumulate these elements.

By applying trace minerals to your soil through the use of compost, you avoid a possible danger of getting too much of these minerals in your food. Scientists have found that some plants, including wheat and barley, have an ability to *concentrate* trace elements. So while a soil may contain five parts per million of lead, certain plants can concentrate it tenfold. Using inorganic sources for some trace elements may tend to overload the soil with these minerals, with the result that too much of a good thing becomes a health hazard. In compost, however, the elements are available in nutritionally-balanced amounts, and this organic *balance* prevents the overloading of one particular element.

When, in 1962, German researcher H. Hick compared the trace element content of manure, city compost, superphosphate, and nitrochalk, he found all but compost to be relatively poor sources of balanced amounts of trace elements. The city compost, however, outran all the others in amounts of copper, manganese, zinc, molybdenum and boron . . . sometimes by as much as 200 to one.

Compost, mulch, leaf mold, natural ground rock fertilizer and ground limestone help provide a complete balanced ration of both major and minor nutrients. Some other good sources are seaweed and fish fertilizers, weeds that bring up minerals from deep in the subsoil, river bottom silt, cover crops with extensive root systems such as alfalfa, and garbage compost and sewage sludge which contains elements from all over the world. Marine deposits such as greensand and oyster shells are especially rich. Besides supplying minerals themselves, these materials release acids upon decomposing that react with the elements in the soil to form compounds plants can use nutritionally.

Basic slag is an industrial by-product, resulting when iron ore is smelted to form pig iron. Rich in calcium, slag also includes valuable trace elements as boron, sodium, molybdenum, copper, zinc, magnesium, manganese, and iron. Its efficiency varies with its fineness. Alkaline in action, slag is most effective on moist clays, loams, and on peaty soils deficient in lime. For light soils, tests show that it's better to mix with greensand or granite dust.

The way these elements become available is through chelation, a process by which nutrients are literally pulled out of soil and rocks by certain compounds. Humus is one such compound converting insoluble minerals to available forms.

Many soil fungi normally produce a variety of compounds that behave as chelators. This may well be a major function of the mycorrhizal fungi which act in the role of root hairs for certain trees and other plants. Without their fungus partners, these plants either grow poorly or are unable to develop at all.

While chelators are not new, during recent years some of the chemical companies have been aggressively marketing synthetic chelates. These pose the danger of being so strong as to dissolve too much of the stored-up nutrients in a short time, leaving little for future crops to draw on.

One of the country's top authorities in the field of organic chelates is Dr. Harvey Ashmead of Albion Labs, in Ogden, Utah. He's tested over 200,000 animals to determine which chelates are absorbed by the animals fastest. Similar tests were conducted on plants.

Two lots of corn were planted in white silicon sand which itself contained no trace minerals. Organic zinc chelate was put into one lot, and inorganic zinc sulfate chelate into the other. The corn to which the organic mineral was added germinated faster and grew taller than that planted in the inorganic chelate. When iron was tried, the organic iron chelate from fish meal scored best, while inorganic iron sulfate prevented growth.

Next, grains were planted. The same proportion of organic and inorganic metals were added to two separate lots of grain, and the roots and stems were tested for mineral content. Once again the organic minerals were absorbed far better than the identical but inorganic ones.

Results Confirm Superiority of Organic Over Inorganic Metals

Having confirmed the superiority of organic over inorganic metals in a series of tests with plants, Dr. Ashmead next applied the same procedures to animals. In one test for absorption, he chelated copper, magnesium, iron and zinc organically with fish meal, soybean meal and whey. These organic chelates were tested in the intestines of rats which showed no signs of mineral deficiency. The inorganic carbonates, sulfates and oxides of the same metals were also tested. The results showed that, under normal conditions, the organic trace minerals were absorbed far better than the inorganic.

The experiments with specially-protected plants and laboratory animals proved organic minerals more effective than inorganic ones when absorbed by living tissues. But laboratory findings become meaningful only when they can be applied to present needs so Dr. Ashmead checked his findings successfully in a far-ranging experiment involving animal nutrition—one of the immediate problems presented by our minerally-depleted environment.

Today most baby pigs are anemic when they are born, and must be given supplemental iron, or they usually die. The hemoglobin level averages 5 gm. percent. However, when pregnant sow pigs were fed organic chelated iron, zinc, copper, cobalt, and magnesium 30 days before they gave birth, their piglets were born with hemoglobin levels averaging 11 gm. percent. Pigs fed the same proportions of inorganic metals delivered piglets of the same weight, but the baby pigs whose hemoglobin levels ranged from 5 to 9 gm. percent, would have died if they hadn't received ad-

Wood chips supply important trace elements to the compost heap.

SOME IMPORTANT TRACE ELEMENTS AND THEIR SOURCES

The following chart includes some important trace elements, their sources, and also the accumulator plants. When making compost, remember that a great diversity of materials used will achieve a more balanced supply of nutrients.

BORON: Granite dust, vetch, sweet clover, muskmelon leaves.

COBALT: Manure, mineral rocks, tankage, yeast, legumes, vetch, peach tree refuse, Kentucky bluegrass.

COPPER: Wood shavings, sawdust, redtop, bromegrass, spinach, tobacco, Kentucky bluegrass, dandelions.

IRON: Seaweed, most weeds. Is usually available for plants in acid, organic soils; the slight acidity dissolves and chelates iron. Humus is one of the best iron chelators known, so compost should help get iron to your plants.

MANGANESE: Manure, seaweed, sea water, forest leaf mold (especially hickory and white oak), alfalfa, carrot tops, redtop, bromegrass. Mulching and applying ground limestone will reduce the poisonous effect of soils containing too much manganese.

MAGNESIUM: Dolomite, high magnesium limestone, magnesite, silicate minerals, soluble salts, lake and well brines, sea water. Add one pound ground magnesium stone, or one quart of sea water to every 100 pounds of compost. Since magnesium is at the core of every chlorophyll molecule, all green matter added to the compost heap is an abundant source of magnesium.

MOLYBDENUM: Cornstalks, vetch, ragweed, horsetail, poplar and hickory leaves, peach tree clippings. For deficiencies, experts recommend raising the pH of very acid soils to 7 with ground limestone.

ZINC: Rock phosphate, ragweed cornstalks, vetch, horsetail, poplar and hickory leaves, peach tree twigs, alfalfa.

> NOTE: Agricultural frit, a type of ground, glassy material that releases trace minerals into the soil slowly, is very abundant in all the elements listed above, except magnesium and iron. It can be obtained as an ingredient in some natural fertilizers.

ditional iron injections. On the other hand, the pigs born from organically-fed mothers didn't require additional iron to keep them alive until they could feed themselves 14 days later.

Dr. Ashmead said, "At weaning, the dramatic results of our organic chelated minerals were truly seen. The average weight of the piglets coming from the treated mothers was 2.65 pounds more than those coming from the control (inorganically-fed) mothers. Thus not only was baby pig anemia prevented by feeding sows our organic chelated minerals, but we got the extra bonus of increased weight gains."

Further evidence of the ability of living tissue to assimilate organic minerals with greater ease than inorganic ones comes from a director of a Utah State fishery division. He reports a high ratio of one pound of fish for every 1.36 pounds of feed containing organic zinc. He added that the treated fish grew more rapidly than those which were untreated, and "their texture approximated the texture of fish in a natural environment."

If production and quality are to be improved, balanced nutrition must be maintained. All essential elements must be present and available in adequate amounts. If a copper shortage exists, increased use of chemical nitrogen fertilizer isn't only useless, it makes things worse. Wrongful fertilization, says Karl Schutte, author of *The Biology of Trace Elements,* "is as little in the interest of the fertilizer industry as it is in the farmer's. It's up to the fertilizer industry to warn their customers against the abuse of their products. It's in their own interest to recognize that to sell a maximum of fertilizer irrespective of its effects is both anti-social and ultimately very bad business."

The organic gardener, careful to supply balanced nutrition to his plants, not only raises food without chemicals, pesticides, or processing, but reaps the personal benefits of food rich with the metallic ions that allow living cells to function properly, bringing glowing good health.

—Jeff Cox

Seaweed—High in Potash & Trace Elements

The use of seaweed as a fertilizer dates back many centuries. As long ago as 1681, a Royal decree regulated the conditions under which seaweed could be collected on the coast of France; the kinds that might be collected were specified as was the manner in which they should be used.

Seaweeds are always associated with a rocky formation of the seabed and most weeds are found in shallow water. The land in such situations is usually a thin cover of sand, or a sandy loam, overlying rock and is exposed to the eroding action of winter gales. The use of seaweed over many generations has resulted in a deep black earth which defies the Atlantic gales.

The seaweeds used as fertilizers belong to two main groups which are distinguished by their habitat. These are the brown weeds which grow between high and low water on rocky situations, and those weeds, also brown, which also grow on a rock bottom below low watermark down to a depth of 60 feet. The first group, often called rockweeds, are relatively small plants, but their growth is usually dense and they are easily collected by pulling or cutting from the rocks; 200 pounds per hour can be gathered easily from a good site. While the number of rockweeds is legion, only two need be mentioned since they are very much more common than any other weeds; these are "knobbed wrack" and "bladder wrack." Both these plants are made conspicuous by the air filled vesicles which enable the plant to reach the surface when the tide comes in; in this way the plants receive more light for photosynthesis.

The sublittoral brown weeds, variously known as Laminaria, oarweeds or tangles, are much larger plants and appear to have a root, a stem and a single large leaf; this similarity with land plants is very superficial and, in fact, these various parts do not function like their counterparts in land plants. Weed beds are often dense and growth is usually about 20-30 tons per acre and may approach 50 tons. All these plants have a frond, or leaf, which is replaced annually, usually about May; a new frond develops and the old growth is detached and cast on the beaches. Unfortunately these fronds soon rot and a decaying heap can be most unpleasant to collect. Stormy weather is usually associated with the fall and winter; during such storms, plants are broken and torn from the sea bed and cast high and dry by the storm. Thousands of tons may be cast on a beach by a single storm. The complete plant is much more resistant to decomposition and such heaps can be removed without qualms.

These two types are very widespread and flourish over most of the Northern Hemisphere. A third type, the giant kelp, grows off the coast of California. These are brown weeds and both annual and perennial types abound; their chief feature is their length which may exceed 200 feet. They shed their fronds between April and June and decay rapidly if the water temperature exceeds 76 degrees F. Like other brown weeds they are cast by storms, and drift weed may also be collected.

Seaweed, a concentrated form of mineral matter from the ocean, has been used as fertilizer for many centuries.

Lastly, some mention must be made of the "gulf weed" which gave notoriety to the Sargasso Sea. Although modern evidence refutes the tales of the medieval mariners, Sargassum species do exist, and cast weed is available on the beaches of the Gulf States.

The chemical constitution of seaweeds is markedly different from that of the land plants. In general, land plants owe their rigidity to cellulose whereas seaweeds contain only about 5 per cent cellulose and owe their mechanical strength to alginic acid. The food reserve of land plants is starch, that of marine plants is laminarin, while the place of sugar is taken by mannitol which is rarely found in land plants. Alginic acid, laminarin and fucoidin are found only in seaweeds.

A typical analysis of a commercial sample of rockweed is:—

Moisture	13.5%
Fats	1.4%
Protein	10.5%
Carbohydrates	51.9%
Fibre	6.1%
Minerals	16.6%

The mineral matter contained:—

Silica	1.48%
Copper	0.002%
Manganese	0.016%
Iron	0.01%
Sulphates	4.18%
Chlorine	1.72%
Magnesia	1.01%
Calcium	1.54%
Potassium	2.83%
Iodine	0.17%
Salt	3.65%
Bromine	0.68%
Aluminum	0.22%
Phosphorus	0.18%

The most interesting feature of seaweeds is their ability to concentrate the mineral matter in the sea, which results in the presence of substantial amounts of a large number of elements in all seaweeds.

Traces of arsenic, boron, cobalt, molybdenum and vanadium are also present. Silver, barium, chromium, lithium, nickel, lead, rubidium, strontium, tin and zinc have also been shown to be present in seaweed. The arsenic content of seaweed (13 parts per million) is less than the normal arsenic content of soil. The chloropyhll content (0.25 per cent) of seaweed is almost as high as that of lucerne and vitamins A, B, E, F, G, & K are present in a sample of the giant kelp. Recent work has shown that vitamin B12 is also present in seaweed and this is the first known plant source of this vitamin.

Seaweeds vary widely in composition from month to month and a typical analysis of a brown weed—*Laminaria cloustoni*—illustrate the extremes of this variation:—

	Minimum	*Maximum*
Alginic acid	17.4%	25.3%
Mannitol	5.3%	17.3%
Iodine	0.6%	0.9%
Fucosterol	0.1%	0.1%
Minerals	25.2%	43.7%
Landnarin	nil	14.1%
Cellulose	4.3%	7.9%
Fucoidin	4.8%	7.6%
Protein	9.6%	12.7%

Such variations are common to all seaweeds and, in this respect seaweed varies more than other fertilizers; but as ordinarily obtained in the wet state, it contains slightly more nitrogen than stable manure, twice as much potash but only about half as much phosphates. A citation of the constituents of seaweed reveals a rich source of trace elements, vitamins and potash, and an ample source of nitrogen. A rather special constituent is alginic acid, whose sodium salt has been shown to be active as a soil conditioner. An organic manure containing 25 per cent of a soil conditioner along with mineral nutrients, trace elements and vitamins can be expected to be a good fertilizer; centuries of usage prove it beyond doubt.

Alginic acid is very susceptible to bacterial attack; this reduces its value as a soil conditioner, but it also means that the microbiological soil flora is correspondingly enriched. This property is undoubtedly the basis of seaweed's activity in a compost heap. When only limited quantities are available, the best use is as a compost accelerator. When chopped, it rots readily, and a 200 pound heap will heat to 100 degrees F. in two days—chopping is essential; otherwise decomposition is slow. Such a heap should be made to form the core of a compost heap which will be ready for use in six weeks. When only small quantities are available, the weed should be chopped and soaked in hot water (one gallon to two pounds of weed) and the mixture poured over the compost heap after soaking overnight. With dry milled weed a temperature of 140 degrees F. is sufficient, but fresh, chopped weed should be scalded and soaked initially at 160–180 degrees F. Such extracts are good compost accelerators and are in commercial use in this country.

In the garden, potatoes should have first claim on seaweed. The sets should be laid in a trench and surrounded with two pounds of chopped weed (equivalent to four ounces of dry weed) ; the trench should be filled in and cultivation completed in the usual manner. If sufficient weed is available, a dressing of half this quantity can be applied before ridging. Root crops and brassicas respond well to seaweed and require the ground to be prepared well in advance and dressed with 100 pounds of chopped weed. Celery and peas should be set in prepared trenches to which a similar dressing is applied and raked into the top few inches prior to planting. Salad crops and onions should receive heavy dressings at 200 pounds or more per 100 sq. ft., while seed beds and borders should have only light dressings of 50 pounds per 100 sq. ft. and should be lightly raked in during the early spring.

Soft fruits, such as gooseberries, black currants, etc., are particularly responsive to seaweed, probably because of its high potash content, and should receive two to five pounds per plant according to their size. Apples and pears may be treated according to size with up to ten pounds for an established standard tree in full bearing.

Herbaceous borders, flower beds and roses should receive 50 pounds per 100 sq. ft., which should be raked in lightly during early spring. Lawns are best treated with finely chopped weed in the fall when the lawn should be spiked with a hollow-pronged fork and covered with half an inch of chopped seaweed. Any slight residue should be swept off with a broom of twigs, prior to mowing in the following spring. This treatment adds vital organic matter to the soil and helps the grass to withstand summer drought. Spring feeding of lawns must be confined to broadcast applications of dried milled weed which should be used at two lbs. per 100 sq. ft. It is essential to avoid adding the weed near the

Molasses residue has been found to be a rich source of potassium.

drip line of trees since this will give a slippery patch; higher feed rates should also be avoided, especially in wet weather, for the same reason. Generally speaking, spring feeding of lawns is not nearly as beneficial as autumn feeding, and should be used only when considerations of time have prevented treatment in the fall.

When a May cast of frond is available, it should be used as a mulch, and spread particularly over shallow-rooting plants such as raspberries and roses, and any of the many garden subjects that wilt in hot sun-baked soil.

Supplies of fresh seaweed are naturally restricted to people living within easy access to a rocky coast, but those who make the occasional weekend trip to the sea will find pleasure and profit in gathering a few sackfuls of weed which can be carried in the car trunk. The rockweeds are best for casual collection since they are superficially dry and can be gathered at all times. Dried, milled weed is also available and 100 pounds of such weed is equivalent to 400 pounds of fresh weed. Dry weed can be substituted for fresh weed and is a good compost accelerator but, of course, it cannot be composted as successfully as fresh weed.

Where an abundant supply of weed is available, it should be collected and spread over the ground to a depth of several inches in the fall and disced in the spring: this is the traditional method in Scotland, where thousands of acres are treated annually in this fashion.

Your local organic fertilizer dealer may carry seaweed fertilizers. If he doesn't, you can order from one of the following companies.

**Bonda Meal and Oil Ltd., Yarmouth, Nova Scotia, Canada. Rockweed, $140 per ton, f.o.b. Yarmouth. ½ ton minimum.*

**F. S. Brisbois, Fonda, Iowa 50540. Algit (Norwegian Kelp Meal), feed additive and fertilizer. Seaborn (liquid, granular, powder), fertilizer.*

**Atlantic and Pacific Research, Inc., PO Box 14366, North Palm Beach, Florida 33403. Made from blended seaweeds.*

**Sea-born Corp., 3421 North Central Ave., Chicago,*

Illinois 60634. Seaborn (liquid and granular), soil and plant food. Algit (Norwegian Kelp Meal), animal food supplement.

**Sea-Organic Corp., 6 Lynde St., Salem, Mass. 01970. Sea-quill, $8.95 for 25 pounds, express paid. Seaweed Meal —20 pounds, $8 express paid.*

**Science Products Company. Inc., 2640 N. Greenview Ave., Chicago, Illinois 60614. Liquid Sea-Weed (Norwegian seaweed). 8 oz. for $1.29, 1 quart for $3.50.*

The Living Soil

Chapter 6

Not so many years ago gardening and farming organically was considered just a fad by those who saw the soil as merely a supporter for plant life. In today's ecologically-aware world, more and more gardeners and farmers know that the soil itself is a living, breathing organism, teeming with life that must be nurtured the natural way—organically—to constantly replenish what it has given us through the harvested crops, and not further depleted and eventually destroyed with chemicals.

J. I. Rodale held a life-long respect for the living soil and taught that the organic gardener's first duty and primary effort should be directed toward maintaining its balance in the way that nature intended. He wrote:

"In the soil there is a whole universe, an amazing but fascinating little world peopled by diverse, multipurposed microorganisms. It is a regular *Alice in Wonderland,* because things are done there by different rote and rule than we are accustomed to.

"In that world you will find all kinds of elements—police and thieves, builders and wreckers, scavengers, magicians and circus performers. As a scientific writer put it, "there you will find complex associative and antagonistic interrelationships." It is the way of life in general. There must be positive and negative, angel and devil, republican and democrat. There must be, in nature's scheme, bacteria which are ready to cause disease if her rules are broken, and there must be bacteria which prevent disease if you give them half a chance.

"As the farmer or gardener works his soil he must be cognizant that it is full of microorganisms. Good tilth and a workable soil structure are the result of the activity of bacteria which exude a gummy substance, a mucus that binds the soil particles together, gluing fine particles into large masses in such a manner as to give it that exquisite quality which you feel when you run your fingers through good earth. In a soil low in bacteria the particles will not "aggregate" as effectively. This good structure of the soil, caused by the activities of bacteria, prevents its washing away by rain. It reduces soil erosion. When the farmer plants his seed he must be aware that without the action of bacteria the roots could not feed properly and he would not secure satisfactory harvest.

"But one of the most important functions of bacteria is to break down organic matter. When a crop has completed its task of growing, the bacteria go to work on the old roots which are left in the ground, decomposing them completely, and transforming them into food for the next set of roots which will take their place. The question may be asked, why do not the bacteria attack live, growing roots? Are there electrified fences around them which scare them off? The answer is no. But the way it works is this. There are hundreds of different kinds of organisms in the soil—each for its own specific purpose. For example there is one that can only extract nitrogen from the air. Certain bacteria are sulfur working organisms. They are purple in color and can only attack sulfur. There are nitrate working, cellulose destroying, sugar and starch working bacteria. It is like a drama—each actor waiting for his cue to come upon the stage to do his bit, and to retire.

"There are certain bacteria whose only function it is to break down dead matter. Anything alive is distasteful to them. The saprophytic bacteria decompose the old roots into substances which are worked over by other organisms and transformed into food for the plants. One of the important functions of soil bacteria, then, is to provide plants with foods, and if the farmer and gardener can so regulate his methods as to consider the well-being of the micro-organisms of the soil, so that they can multiply to abundance, he will be well repaid for his efforts. To illustrate one aspect of the advantage of a bacteria-rich fertility, when a crop of corn is harvested and the heavy stalks plowed under, if there is a lack of bacteria, a temporary indigestion will occur that will reduce the yield of the next crop.

"We had an interesting experience in the winter of 1941

which showed us how valuable bacteria could be on a farm. We were fattening steers purchased at the Lancaster stockyards. We did not maintain a pasture but fed them in the barn, giving them access to an open corral. In the barn we maintained a heavy bed or litter of straw, and the practice was to clean it out every two or three weeks. But it became difficult to find the help to do all the work, because of war shortages. We could not find the time to clean out the litter, and therefore had to resort to what is called the English fold-yard method.

"In this system, instead of cleaning out the litter, we kept adding more straw on top of it every week. The action by the steers, trampling it down, did not permit it to "hit the ceiling." We were surprised to observe composting of a sort taking place. The litter, moistened by the urine of the animals, and maintained at the proper temperature for the working of bacteria, entered into the beginning processes of decomposition, working from the bottom up. By spring, when the barn was cleaned, the litter was a quite brown material and was composted enough so that it could safely be spread on the land for use by a crop which was to be planted a few weeks later.

"We followed the fold-yard method again the next winter, 1942, but this time noticed a singular difference. In the preceding season the characteristic manure odor had been quite noticeable, but this time there was practically no detectable smell. Here was something extremely significant which must be given some thought. It could not be ignored. The facts soon became clear. What had happened was that the composting process had bred billions of bacteria, fungi and the other microorganisms that take part in the process of decomposition. They had multiplied to such an extent, and they had saturated the earth floor so thoroughly that it had become a rich culture-bed of the kind of bacteria that break down organic matter.

"When the new droppings came in contact with the ground in the second year, washed down by the animal urine, billions of microorganisms were there to penetrate it and to begin the work of decomposition. It was the same as in the making of a compost heap. A properly made compost will not give off an odor if conditions have been made ideal for the bacteria and fungi which take part in the process of fermentation.

"This same season I attended a farm auction where I could not help overhearing the conversation between two farmers. One of them made the statement that the litter in his barn was kept free of odor. When I expressed an interest in hearing more about it he revealed the fact that he did it with a chemical that he bought from Sears Roebuck. Later, I checked on it and found that it was a bacteria-killing antiseptic, something like Lysol.

"This experience gave me a graphic illustration of the difference between the organic and chemical methods in farming. This farmer was reducing his odor by killing bacteria while we were doing it by multiplying them. This man, therefore, was applying a poorer manure to his land. In our farming we give a thought to maintaining a biologic state of health in the soil.

"Bacteria, if understood properly, and harnessed in the right way by the farmer can put money into his pocket. They can inactivate larvae of dangerous disease-producing insects, by getting at them at the source—in the soil. They can be a factor in getting more food for crops from organic and inorganic stores of it in the soil.

"There is one aspect of the little micro-biological world in the soil which is amazing. It carries its own police force. You have heard of penicillin. Perhaps you have had the use of it. Well, penicillin is an organic substance excreted by the penicillium mold, which is a microorganism closely related to bacteria, although usually somewhat larger in size. There is a large family of anti-biotic organisms, little police, of which penicillin is only one. There is streptomycin, terramycin and many others in this group, which have been in the soil since time began. They are a protective device of nature—little policemen that are in the soil to keep the disease-producing bandit organisms from getting too many ideas. They lull them and keep them inactivated by excretions from their bodies.

"We know from experience on farms where conditions in the soil are kept as natural as possible, withholding therefrom all strong chemicals and poison sprays, that there will be few signs of disease in plants, thanks to these anti-biotic organisms. Under these conditions this type of micro-organism is in good supply, in vigorous health, and in a position to regulate competently the life that goes on below ground.

"Now, suppose you were a well-behaved law-abiding little bacterium, down in the dark earth, minding your own business, doing your work like a good little bacterium—turning nitrite to nitrate, or hydrogen sulphite to elementary sulphur, and suddenly what seems like a bucket of poisonous chemicals is poured over your head. Let's say you are one of the soil's little policemen, a penicillium with eight stripes for special bravery in action, and your farmer sprays you. How would you feel, if you had any feelings left to feel with? What kind of civilization destroys its policemen? How long could such a culture endure?

"Let us say you're a soldier and a gas attack occurs, but you have no gas-mask. The Government people might say, 'When we have no more of our own soldiers we will hire Hessians.' That is exactly what the Government agricultural scientists are saying. They are killing off the wonderful little soldiers in the soil, with a feeling that they can hire Hessians—or chemical fertilizers to do the work, but what a disillusioning they are due for. I can tell you that the more chemical fertilizers used the more Hessian flies will be found in the wheat and the lower the number of bushels produced per acre. The scientist has his head in the ground, like an ostrich, but does not see the bacteria with which it is saturated.

"Even if you are only a gardener, be aware that all of this knowledge applies to you, also, to your soil, to your radishes and to your begonias. By nursing along and encouraging the microorganisms in the soil, by feeding them sufficient of the kind of food they prefer, namely, living organic matter, shunning strong chemicals in any form, you will develop a much finer soil, and your results will be wonderful. Besides, in the beginning as well as in the long run it is cheaper."

—J. I. Rodale

Make Your Own Organic Fertilizer

Chapter 7

Before you mix your fertilizer you'll want to know as much as possible about what your particular soil needs. There are two ways to find out what nutrients your soil is hungry for. You can send a sample of your soil to a laboratory or to your state college, or you can buy a testing kit and make many of the necessary tests yourself. It doesn't hurt to use both methods, because you will be able to double check your results. A home testing kit is valuable because it enables you to make frequent periodic tests of your soil. Most people don't realize that the nutrient supply in the soil varies greatly from one season to another.

A soil test will tell you what nutrients are "available" or soluble in your soil. There are also "unavailable" or insoluble nutrients in your soil which plants can feed on, but the supply of soluble and insoluble nutrients is usually quite similar. If your soil is low in soluble phosphorus, it will probably be low in insoluble phosphorus, etc.

The best place to start your soil analysis is with its pH. Is it alkaline or acid or neutral? This is important, because correcting the pH will often release supplies of major and minor plant foods. A soil that is too acid will not release "unavailable" nutrients properly, and the same is true of a soil that is too alkaline.

Soil Samples: Collecting the soil sample is the first step in making a test. In fact, collecting the sample could be considered the most important and critical part of the testing procedure. There is often a great variation in soil condition nutrients in various parts of a garden or field, so it is important to collect a number of samples from different locations. These samples can be tested individually and the results averaged, or you can mix all the samples together and use a portion of this "homogenized" sample for your test. Always be careful to make sure that your collecting shovel or container is not contaminated by a fertilizer. That might throw off the results.

The test is made by putting a small portion of the sample in a test tube and then introducing one or two "reagents." A reagent is a chemical which reacts with the nutrient being tested and shows the quantity of the nutrient available by changing color. Color charts are supplied with the test kits, and the final analysis is made by checking the color of the solution in the tube with the test chart for the nutrient being tested.

The first and foremost rule in mixing organic fertilizers is KEEP IT SIMPLE. Don't develop a "numbers" complex over fertilizer formulas. Here's a valuable guide for choosing and using your organic materials.

Fertilizing garden soil with organic materials is a simple job—except perhaps for the person doing it for the first time. For example, take the case of Mr. J. D., an organic gardener from Louisiana. He's just had his soil tested by the State University and was advised to use 800 pounds of 8-8-8 per acre. He writes: "Since I'm gardening the organic

ORGANIC FERTILIZERS

—replenish the soil.
—keep soil friable.
—promote beneficial soil life.
—prevent hardpan soil conditions.
—increase crop yields
—produce crops with the proper nutrients and that old-time flavor.
—are safe.
—maintain a natural balance in the soil, resulting in pollution-free water—water you can drink without the fear of nitrate poisoning.
—protect certain crops from disease.
—benefit the environment by recycling wastes.
—are generally neutral in reaction and non-corrosive.
—are inexpensive.
—are easy to use.
—grow larger plants.

way, and don't want to use commercial fertilizers, I thought of the following substitutes that I can easily get: cottonseed meal (7 per cent nitrogen, 2–3 per cent phosphorus, 1.5 per cent potash), phosphate rock (30 per cent P) and wood ashes (1.5 per cent P, 7 per cent K). How many pounds of each should I apply to equal the fertilizer amount in 8-8-8?"

Like so many other organic growers, Mr. D. has a good, clear idea of what materials to use and where to get them. His main trouble, as we see it though, is that he's suffering from a "number" complex. Because some expert told him to use 8-8-8, he's going to struggle to match it—only with organic fertilizers instead of chemical. Aside from the needless mental activity (perhaps even anguish) caused by this

CHEMICAL FERTILIZERS

—pollute the nation's lakes, rivers, and streams. Dr. Barry Commoner reports that pollution by nitrates from inorganic fertilizers equals pollution by sewage. Chemical fertilizers encourage the growth of "algae blooms" which make "huge cesspools of the nation's lakes," says Commoner.
—have caused numerous cases of nitrate poisoning, methemoglobenemia; over 3,000 head of cattle and 11 persons have been killed thus far.
—increase the infant mortality rate among female babies, says Dr. Abraham Gelperin of the University of Illinois.
—are reported by Japanese scientists to contain the same type of carcinogens as nicotine.
—deteriorate soil friability.
—create hardpan soil.
—destroy beneficial soil life, including earthworms.
—alter vitamin and protein contents of certain crops.
—make certain crops vulnerable to disease.
—prevent some plants from absorbing some needed minerals.
—produce foods that just don't taste as good.
—are costly.
—are a leading cause of blindness on the farm.
—are corrosive to farm equipment and storage areas.

complex, there are other disadvantages:

First, it's often difficult to equate the organic ratio with the chemical one. The result may be that the "new" organic gardener may say that the organic method is difficult or confusing.

Secondly, and perhaps the most common trouble, a lot of gardeners and farmers make the big mistake of not using organic fertilizers heavy enough on their first applications. We've found this to be true time and again. Advertisements of chemical companies who have just come out with an expensive fancy mix or super blend may advise applying at the rate of 200 pounds per acre, or a pound or two for the whole vegetable garden. Then when the organic grower wants to convert, he still thinks of such applications.

Fertilizer Program for You

That's what harm the "number" complex can do in terms of worry, confusion and inefficiency. Our advice to Mr. D. of Louisiana (and every organic gardener and farmer worrying about chemical ratios) is to forget about the numbers, and concentrate on a long-range fertilizing program. Once you do this, you'll find yourself growing better plants . . . and sleeping better, too.

Soil tests are a valuable guide to all gardeners and farmers, as they will indicate when and how to increase fertilizer applications. At first, these tests will show what elements are needed most; future tests will tell how well your fertilizing program is working out. But when these results are accomplished by suggestions for chemical fertilizers, don't feel you *must* come up with the *exact* mixture in organic fertilizers.

Disregard the Numbers

When you want to add nitrogen to your soil, take your pick of tankage, manure (fresh or dried), homemade or commercial compost, sludge, any of the vegetable meals as cottonseed, linseed, soybean or peanut. Because of the high nitrogen content of blood meal and dried blood (12 to 15 per cent), use these materials more sparingly.

For phosphate, you have your choice of rock or colloidal phosphate, bone meal, and some materials named above as dried blood, the meals, and manure. Natural potash sources include granite dust, greensand marl, wood ashes, cocoa shells, kelp and many plant residues.

All these materials can be worked into the soil in spring or fall, top-dressed around growing plants, used as a mulch or added to the compost heap. Keeping a high humus content in your soil means that many of these fertilizers will be more readily available and will remain available.

—Jerome Olds & Fred Veith

A Successful Organic "Recipe"

Paul Mahan of Hobe Sound, Florida, starts in September by spreading about two inches of compost over the entire plot and digging it into the soil. "At planting time, I mix together—in the wheelbarrow—some well-rotted cow manure, ground phosphate rock, granite dust, commercial all-organic fertilizer, dolomite, bone meal and ground tobacco stems. (Remember, exact amounts are not important, and I substitute one all-organic mix for another, depending on what's available.)

"I then open the furrows about 6 inches deep and dig in a generous amount of this mixture. I level the soil and mark off the rows to the planting depth required by the seed. This method puts the plant food under the roots, encouraging deep rooting and reducing irrigation needs.

The seed is covered with peat moss, which I've found aids quick germination and eliminates damp-off problems. I have no idea what the pH of the soil is, but why worry when everything grows? When the plants are coming up, I water once or twice with fish emulsion.

"This year, since I couldn't find any hay, I mulched with sugar cane litter—made out at the sugar mill and available at the garden store. It costs $3.25 per bale. Two bales did my garden, and I believe it will prove to be better than hay.

"The proof that plants respond to treatment is that I picked kohlrabi in 35 days, snap beans in 45, tomatoes in 60 and all vegetables ahead of the time as stated in seed catalogs."

"My fruit trees get their heaviest feeding in the fall. I have a large compost bin and keep adding to it all year. I don't turn it, but just let it compost and add to the top as it falls. A two-inch application is given the trees in September.

"I use about the same organic materials to feed the trees as the garden. Spread around the trunks and out to the drip line, the soil conditioners soon become part of the soil.

"Sometime in August, I buy a load of heavy soil (called 'topsoil' by the nursery), mix it with a load of cow manure, and cover with black plastic to compost for about two weeks. The minerals (bone meal, rock phosphate, dolomite, etc.) are added. This combination is spread under the trees about one-half to one inch deep. Between fruiting periods, I spray the trees with fish fertilizer.

"Some trees that fruit more than once a year get an extra application of organic blends like Fertrell and Hybrotite. The loquat fruits twice a year, and my carambola had 4 crops last year.

"I do not mulch my citrus fruit trees because they are very shallow-rooted. If mulched, the feeder roots will come to the surface of the ground. Lychee trees like lots of mulch and manures. The carambola must have an acid soil. To keep it healthy, I mulch it with oak leaves and oak leaf-mold.

"In all, I have some 30 different varieties of fruit trees growing here—including mangoes, limes, cherries, plums, avocados and figs—and they all do well on the organic feeding I cook up for them."

40 GREAT ORGANIC FERTILIZERS AND SOIL BUILDERS

Compost — probably the best organic fertilizer
Sludge — safe and easy to obtain
Basic slag — loaded with calcium and trace elements
Blood meal and dried blood — one of the best sources of nitrogen
Cottonseed meal — low pH makes it good for acid-loving crops
Grass clippings — 1 lb. of N and 2 lbs. of potash in every 100 lbs. of clippings
Greensand — 6–7 % potash
Granite dust — a highly recommended source of potash
Vegetable residues — are attractive and easy to use
— *Cocoa beans* — *Oats*
— *Buckwheat* — *Rice*
— *Cottonseed*
Leaf mold — nitrogen content as high as 5%
Leaves — abundant source of humus and mineral material
Fresh manure — a basic natural fertilizer for centuries
Dried manure — always useful in the garden
Peat moss — aerates the soil and improves drainage
Phosphate rock — a real time-saver
Colloidal phosphate — an excellent source of phosphorus and trace elements
Sawdust — when well rotted, an excellent fertilizer
Seaweed and kelp — both are high in potash and trace elements
Tankage — packed with phosphorus
Wood ashes — alkaline, great for mixing with other fertilizers
Wood chips — higher nutrient content than sawdust
Soybean meal — a very useful soil builder
Castor pomace — available at most garden centers
Tobacco stems — useful and inexpensive
Hay — fairly rich in nitrogen
Straw — will easily fit into your fertilizer program
Hoof and horn meal — loaded with nitrogen
Dry fish scraps — where obtainable, a very good source of phosphorus
Brewery and cannery wastes — nitrogen-rich
Feathers — an excellent soil builder
Garbage — safe and ecologically sound
Coffee grounds — a household supply of nitrogen
Lobster shells — a potential source of nitrogen
Molasses residue — extremely rich in potassium

Sludge, Garbage and Leaves

Chapter 8

How good is sewage sludge as a fertilizer and soil conditioner? Is it safe to use? What is its nitrogen and phosphorus content? Where can you get it?

To come up with some of the answers, we wrote to state sanitary engineers throughout the United States to find out what cities have been marketing sludge or at least making it available for gardeners and farmers to use. Next we wrote to the superintendents of the plants themselves for more detailed answers. Finally we examined the extensive research already done on the subject.

The use of sewage sludge by gardeners and farmers throughout the United States has been climbing upward in recent years. In cities where sludge is sold, such as Boise, Chicago, Wichita, Grand Rapids, Duluth, Omaha, Santa Fe, Schenectady, Houston, Roanoke and Milwaukee, superintendents of sewage treatment plants report that demand for sludge has been increasing.

Most cities and towns, probably the one you live in, don't sell their sludge, but make it available to local gardeners and farmers free at the plant site. In Illinois, for example, according to Sanitary Engineer Carl Gross, "virtually every city and sanitary district which operates a sewage treatment works makes the digested and air-dried sludge available at the treatment works to anyone. In a few instances, there is a nominal charge, but in general, the sludge is available at no cost."

Activated or Digested?

The fertilizer value of the sludge produced depends largely on which processing method is used.

1. Activated sludge: This kind is produced when the sewage is agitated by air rapidly bubbling through it. Certain types of very active bacteria coagulate the organic mat-

ter, which settles out, leaving a clear liquid that can be discharged into streams and rivers with a minimum amount of pollution.

Generally, activated sludge is heat-treated before being made available to gardeners and farmers; its nitrogen content is between five and six per cent, phosphorus from three to six per cent. Its plant food value is similar to cottonseed meal—a highly recommended organic fertilizer.

2. Digested sludge: This type of residue is formed when the sewage is allowed to settle (and liquid to drain off) by gravity without being agitated by air. The conventional anaerobic digestion system takes about 10 to 14 days from the time the sewage reaches the sedimentation tank until the digested solids are pumped into filter beds, often sand and gravel, for drying. The final step is removal of the dry material, either to be incinerated or used for soil improvement.

A few medium-sized cities located in agricultural areas dispose of a part of their short-time activated and subsequently digested sludge in liquid form. This is delivered to farmers within a radius of about ten miles and is used for direct application to land.

Digested sludge has about the same fertilizer value as barnyard manure. Nitrogen varies from two to three per cent, phosphorus averaging about two per cent. It often has an offensive odor that persists for some time after application to a soil surface during cool weather. "This odor differs greatly in character, however, from that of raw sludge," Dr. Myron Anderson, senior chemist for the Agricultural Research Service at Beltsville, Maryland, states, "since drastic changes have taken place during digestion. The odor from digested sludges may be eliminated by storage in a heap during warm weather."

This talk of odor reminds me of the experience one editor of a trade publication in the public works field recently told me about. He had moved into a new home on Long Island and had the most common problem of how to build a good, permanent lawn. Because he knew about the value of sludge, he contacted the superintendent of a local disposal plant and arranged to have him deliver several truckloads of sludge.

The result—the most luxuriant lawn in his neighborhood. Side result—a next door neighbor offended about the odor.

Several months later, this same neighbor wanted to know whom to contact so he could have some sludge spread on his lawn. That's how the story ended—green lawn winning out over temporary smell.

Amounts to Use

The city of Marion, Indiana, has printed an excellent little pamphlet which includes the following suggestions for using sludge as a fertilizer and soil builder:

(1) *For starting new lawns*—prepare the seed bed by mixing the sludge with the soil. A minimum dosage of one part sludge to two parts soil may be used if the soil is of a heavy clay texture. Spade the ground to a depth of at least six inches, making sure the sludge is thoroughly intermixed with the soil. Never have layers of sludge and layers of soil in the seed bed.

(2) *For feeding well-established lawns*—the sludge should be applied in the winter and early spring months when the ground is frozen. The lawn may be covered with a layer of sludge one-half inch deep during the cold months of December, January and February. This cover will provide an insulation to protect the grass roots from the harmful effects of freezing and thawing during this time of the year. In addition, ample plant food will be made available for a luxuriant growth of grass in the spring.

(3) *For reconditioning old lawns*—on yards or portions of yards where the soil has been too poor to support an average growth of grass, it is recommended that the procedure under (1) above be followed. It should be remembered that grass roots grow into the soil to a depth of six or seven inches only when the ground is of such a composition that will permit this growth. To top-dress and seed a section of ground that has not been properly prepared is a waste of time and money.

(4) *For gardens and farm land*—sewage sludge provides an economical means of replenishing land with nitrogen, phosphorus, and humus material. It should be put on the

land just previous to the fall or spring plowing and cultivated into the soil. The sludge cake may be applied as a manure to farm land at the rate of ten to 15 tons to the acre. The weight of the sludge is approximately one ton to the cubic yard.

Heat-dried activated sludge is generally acid, its pH averaging 5.0 to 6.0. Air-dried digested sludge after a short period of storage will normally show a drop in pH to about 6.0. Experience shows that the continued use of any type of sludge, unless previously conditioned with lime, requires the periodic addition of lime to the soil to prevent harmful acidity.

Sludge Is Safe

In the U.S. Dept. of Agriculture report on sludge, Dr. Anderson writes:

"Activated sludges need heat-treatment before use as fertilizer. Such treatment is normally provided for material to be marketed. The heat used for drying normally accomplishes the destruction of dangerous organisms. *This means that properly heat-dried activated sludges may be used with confidence regarding their safety from a sanitary standpoint . . .*

"It seems that states have generally accepted the conclusions of the Committee on Sewage Disposal of the American Public Health Association that *heat-dried activated sludges are satisfactory from a sanitary standpoint, and that digested sludges are satisfactory except where vegetables are grown to be eaten raw. All danger is thought to be removed by action in the soil after a period of about three months during a growing season."*

What kind of results can you expect when you add sludge to the soil? LeRoy Van Kleeck, Principal Sanitary Engineer of Connecticut's Dept. of Health and an experienced gardener, answers that question quite conclusively in his report, "Digested Sewage Sludge as a Soil Conditioner and Fertilizer":

". . . Sludge is particularly adapted to lawns. Sludge deepens the green color of grass and stimulates a luxurious growth. Its benefits seem noticeable for several years. It should be applied late in March and again if desired early in September.

"The home owner may well use it for the flower garden. Here it provides a much needed humus for the hot summer months as well as a moderate but long-yielding nitrogen.

"Its use is also indicated for trees and shrubs. Trees fertilized with sludge frequently have a healthier foliage, both in amount and color, and retain their leaves for a longer time in the fall than nearby unfertilized trees.

"Soils repeatedly planted to growing crops are greatly benefited by the organic humus supplied by manures, peat moss, green cover crops or sludges. The value of sludge should not be judged solely by comparison of chemical analyses with artificial fertilizers, *but by the results it produces in plant growth.* Dependence exclusively upon commercial fertilizers without consideration for the maintenance of humus content and good soil structure is an unsound practice . . ."

Modern engineering advancements have taken sludge out of the dirty-word classification and have made it a valuable organic material.

The potential is tremendous. Using sewage sludge as a fertilizer and soil conditioner *can* be profitable and practical. We're hopeful that these two reasons are enough to make its use much more widespread in the very near future.

To get some for use on your soil, just call up your City Hall or sewage treatment plant. Many plants deliver truckloads at a very nominal charge, or you can rent or borrow a truck and do your own hauling.

Apply sludge to your lawns, work it in around trees, shrubs and flower beds, use it in the garden, add it to the compost heap. Your results will prove sludge is worth using.

—Jerry Goldstein

Leaves Are Great

Here are some gardeners who have been putting their leaves to work over the years, and benefiting by doing so:

1—Shredding them and then working them into the compost pile;

2—Using them as fertilizer by incorporating them into the soil;

3—Using them as mulch—and very good mulch they make, too—in the planting row, around tree basins and in flower beds;

4—As a straight planting medium for root crops like potatoes.

How Communities Handle Leaves

Towns, boroughs, municipalities handle their leaves in a variety of ways. In Scarsdale, New York, leaves are composted and sold by the bag. Maplewood, New Jersey, also composts its leaves and then invites the residents to come down to City Hall and Toronto, Canada, also composts leaves, and sells the end-product for $1 per bag.

The Town of Brookline, Massachusetts, saved and "thoroughly rotted" its leaves at the request of the Chestnut Hill Garden Club.

In Berlin, Germany, leaves from the city's 200,000-plus trees are either mixed with garbage and composted, or sold to private gardeners and agricultural groups. The price is low—four marks or $1 a truckload. The buyers use the leaves to cover hotbeds, in hothouses and root crop pits, and make compost of them to a small extent.

In Toronto, Canada, king-sized street vacuum cleaners suck up leaves, which are then dumped in long windrows and composted. The following year, the material is shredded, allowed to mature further, and then sifted. The finished product—leaf mold compost—is then bagged and sold at a profit to the general public. It is worthy of mention that the "tailings," or leftover leaf compost, are used in a land-drainage reclamation project.

City residents in Portland, Oregon, are encouraged to order their leaves well ahead of the season so the Public Works Commissioner knows what the demand is and is able to plan his delivery schedule. Response is evidently widespread because the "available supply of leaves is prorated among the applicants" when the leaf crop is not big enough to meet the demand.

That was the problem facing Barbara Gilford when the city truck dumped a full load of leaves on her 60-by-130-foot "bit of land" in Frederick, Maryland. Mrs. Gilford had read in OGF about helping the community solve its leaf problem, and "all that free mulch appealed to me." Now it was up to her to put it to work. The truck driver didn't think much of her chances. "If you need any more, just call," he chuckled and drove off.

So Mrs. Gilford "attacked that mountain with a wheelbarrow and my bare hands. First, I spread a heavy 10-to-12-inch layer around the shrubs. When I was finished, I hadn't even dented the pile."

Then she tried to put the children to work, but they merely discovered that "the pile was great for jumping." As for Mr. Gilford—"he just shook his head—didn't complain, and didn't offer to help." So Mrs. Gilford pitched in over the days, aided entirely by "the two pre-schoolers who were enthusiastic about barrow rides!" But they piled leaves around the raspberries, thickly over the asparagus bed, and down to cover the ground where the rhubarb grew.

There were incidental difficulties like the wind scattering the leaves, taking them from where they were needed and depositing them where they weren't. But finally the day arrived when "the pile inside the gate grew to be four feet high in places, covering the entire area as we planned." And then there was that golden moment when the children coming home from school shouted: "Mommy, where has our leaf pile gone?"

Next year, although she was nervous about cold soil and too thick a leaf cover, Mrs. Gilford had fine results. She applied a load of "lovely, strawy manure" over the leaves, which she moved aside at planting time. "The peas were the best I've ever grown," she reported. "The asparagus and rhubarb made it through those leaves just fine, though they were later than usual."

And then, significantly, she adds: "In fact, the soil stayed cold under the leaf mulch long after my usual planting time—though things seemed to catch up later. I had good head lettuce for the first time. The tomatoes, leaf-mulched, grew well, and so did the peppers. A few squash seeds, planted in a pocket of rich earth in the middle of a thick pile of leaves, grew as fast as Jack's famed beanstalk. The daffodils were tremendous."

Some Lessons Learned

Mrs. Gilford learned a few lessons about leaf-gardening. "The next time I get a truck of leaves—yes, there certainly will be a next time," she wrote—"I'll ask for them early enough, so that spreading won't be a rush job. I may even get up nerve to suggest that the officials might look into composting their leaves and selling them the way Norwalk, Connecticut, and Scarsdale, New York, do. That should take a load off the new landfill, at least.

"When they arrive, I'll put my mountain of leaves through our lawnmower or buy a shredder. I'll plan to pull away the mulch from part of the garden late in February just long enough to let the ground warm up, and get some early crops in. I'll encourage some of the rhubarb and asparagus to put out early growth in the same way."

"We are leafers—perhaps loafers, too—from way back," Jeanne Rindge reported from upstate New York where she and her family produced "vegetables that were always in demand. We never had enough," she recalled, "and many of the boys' clients drove many miles to obtain our corn, squash, tomatoes, beans. They said they could tell the difference in the flavor."

And—what did the Rindge family do to obtain such splendid—and profitable—results.

"We were greedy," she wrote, "and prevailed upon the town fathers to dump their annual truckloads of leaves on our place." Although hauling proved a "burdensome job," rewards were not slow in coming. "As our garden fertility grew through the use of well-balanced farm compost, we spread the leaves directly on the large garden which supplied our boys' flourishing roadside stand, and disked them in at the season's end.

"In the ten years we were on the farm, our steadily improving soil fertility paid off in increasing healthy and tasty farm produce. Our vegetables also became virtually disease-free at a time when our neighbor, across the hedgerow, was spraying nine times a season for bean beetles. In addition, our heavy humus soaked up the spring rains like a sponge, while the hedgerow ran brown with his runoff."

Later the Rindges moved to town, where they put the same principles to work for them on a smaller place. "Leaves," Jeanne wrote, "have come to our rescue under every imaginable situation, and they have never failed us. Leaves can be composted without manure by piling them with alternate layers of topsoil—several inches of leaves and a sprinkling of garden soil. Or, as we have repeatedly found, leaves can be tossed in the corner where, with the aid of soil organisms and earthworms, they soon will be ready to work for you as mulch, fertilizer and soil-conditioner."

By now we hope you're convinced that you're doing yourself a favor when you garden with leaves. And—you owe it to the garden and homestead that you've planned for and are working to sustain and bring to fruition. Finally, you owe—yes, owe it to the larger community, the municipality in which you live, to find out what is being done with the leaves. You'll lighten the budget and ease taxes if you recycle leaves in your garden and around your home grounds.

You'll also have a lovelier place to live in and better food for the table. Using leaves can do all that.

Green Manure Is Tops

The recent development of power garden equipment has opened the door to one of the most helpful practices for the organic gardener—utilizing green-manure crops. Until the innovation of rotary tillers and garden tractors, green-manuring was an awesome, if not impossible task. But now that many gardeners classify power equipment as "indispensable," the practice is becoming more widespread—and for good reason.

Aside from the fact that it is sometimes impossible to compost enough material for some larger gardens, green-manuring saves hours of hard work and retains soil fertility economically.

Green manure plants are one of the best soil conditioners ever discovered. They cost little, take little time to use and provide the answer to good soil tilth.

What green manure plants are best and how to grow them have sometimes been bothersome questions. This article will tell you which green manure plants you can use and how to grow them.

Now is an excellent time to start thinking about green manuring. Acreage taken out of production can soon be planted to soil-building crops. In warm weather green manure plowed into the ground decomposes rapidly. If you start early, you may even be able to use two successive green manure plantings this summer.

Here are some tips for efficient green manuring:

1. Legumes make good green manures, because they increase soil nitrogen supplies.

2. It is good to plant a cash crop within a short time after working in green manure—to take advantage of the nutrients given off by the decaying plant matter.

3. With sandy or porous soils, green manures can be left at or near the surface to retard leaching. In clay soils, it's best to incorporate the green manure into the soil where it will break down faster, since leaching in such soils takes place slowly.

Study the following list to see the plants best suited to your needs:

ALFALFA—deep-rooted perennial legume; grown throughout U.S. Does well in all very sandy, very clayey, acid, or poorly drained soils. Inoculate when growing it for the first time, apply lime if pH is 6 or below, and add phosphate rock. Sow in spring in the North and East, late summer elsewhere, 18 to 20 pounds of seed per acre on a well-prepared seedbed.

ALSIKE CLOVER—biennial legume; grown mostly in the northern states. Prefers fairly heavy, fertile loams, but does better on wet, sour soil than most clovers. Sow 6 to 10 pounds per acre in spring, or may be sown in early fall in the South.

ALYCECLOVER—annual legume; lower South. Prefers sandy or clay loams with good drainage. Sow in late spring, 15 to 20 pounds of scarified seed per acre.

AUSTRIAN WINTER PEA—winter legume in the South, also grown in early spring in the Northwest. Winter-hardy north to Washington, D.C. For culture, see *field pea.*

BARLEY—annual non-legume; grown in the North. Loams, not good on acid or sandy soils. In colder climates, sow winter varieties, elsewhere spring varieties, 2 to 2½ bushels per acre.

BEGGARWEED—annual legume; South, but grows fairly well north to the Great Lakes. Thrives on rich sandy soil, but is not exacting; will grow on moderately acid soils. Inoculate when not grown before. Sow 15 pounds of hulled and scarified seed or 30 pounds of unhulled seed when all danger of frost is past. Volunteers in the South if seed is allowed to mature.

BERSEEM (*Egyptian clover*)—legumes for dry and alkali regions of the Southwest. Usually grown under irrigation. Will not stand severe cold.

BUCKWHEAT—non-legume; grown mostly in the Northeast. Tops for rebuilding poor or acid soils; has an enormous, vigorous root system, and is a fine bee plant. Sow about 2 bushels to the acre, anytime after frost. Can grow 3 crops, 40 tons of green matter per acre, in a season. Uses rock fertilizers very efficiently.

BUR CLOVER—a fine winter legume as far north as Washington, D.C., and on the Pacific Coast. Prefers heavy loams, but will grow on soils too poor for red or crimson clover; if phosphate is supplied. Sow in September, 15 pounds of hulled seed or 3 to 6 bushels of unhulled seed per acre. Volunteers if allowed to set seed.

COW-HORN TURNIP—non-legume; widely adapted. Its value lies in its enormous long roots that die in cold weather and add much organic matter in the spring. Plant in late summer, 2 pounds per acre.

COWPEA—very fast-growing annual legume. Thrives practically anywhere in the U.S. on a wide range of soils. A fine soil builder, its powerful roots crack hardpans. Inoculate when planting it the first time. Sow anytime after the soil is well warmed, broadcasting 80 to 100 pounds or sowing 20 pounds in 3-foot rows.

CRIMSON CLOVER—winter-annual legume; from New Jersey southward. Does well on almost any fairly good soil; on poor soil, grow cowpeas first for a preliminary buildup. Sow 30 to 40 pounds of unhulled seed or 15 to 20 of hulled, about 60 days before the first killing frost. Inoculate if not previously grown. Dixie hard-seeded strain volunteers from year to year in the South.

DALEA (*Wood's clover*)—legume; northern half of U.S. Still being tested, but shows promise for strongly acid, sandy soils. Volunteers for many years.

DOMESTIC RYEGRASS AND ITALIAN RYEGRASS—non-legume; many areas. Wide range of soils. Sow 20 to 25 pounds in the spring in the North, fall in the South.

FIELD BROME GRASS—non-legume; northern half of U.S. Widely adapted as to soils. Good winter cover, hardier than rye. Sow in early spring or late summer, 10 to 15 pounds per acre.

FIELD PEAS—annual legume; wide climatic range. Well-drained sandy to heavy loams. Sow 1½ to 3 bushels, depending on the variety, in early spring in the North, late fall in the South. Inoculate first time grown.

HAIRY INDIGO—summer legume; deep South. Moderately poor sandy soil. Makes very tall, thick stand. Sow in early spring, 6 to 10 pounds broadcast, 3 to 5 drilled.

KUDZU—perennial legume; South to Central states. All but the poorest soils. Commonly allowed to grow for several years before plowing under. Seedlings planted in early spring.

LESPEDEZA—legume; South and as far north as Michigan (Korean and Chinese varieties in the North). All types of soil, but Chinese is particularly good for poor, sour soils—for these, it's one of the best fertility builders available. Sow in spring, 30 to 40 pounds. Benefits from phosphate rock. Inoculate first time grown. Will volunteer if seed is allowed to set.

LUPINE—legume; Southeast to North. Sour, sandy soils. Blue lupine is a fine winter legume in the South; white and yellow are most often grown in the North. Sow in spring in the North, late fall in the South, 50 to 150 pounds, depending on the variety. Always inoculate.

OATS—non-legume; widely grown. Many soils. Winter oats suitable for mild winters only. Sow 2 bushels in the spring.

PEARL MILLET—non-legume; as far north as Maryland. Fair to rich soils. Commonly planted in 4-foot rows, 4 pounds per acre.

PERSIAN CLOVER—winter-annual legume; South and Pacific states. Heavy, moist soils. Sow in the fall, 5 to 8 pounds. Inoculate. Volunteers well.

RAPE—biennial non-legume; many areas. A rapid grower in cool, moist weather. Sow 5 to 6 pounds per acre.

RED CLOVER—biennial legume; practically all areas, but does not like high temperatures, so is most useful in the North. Any well-drained fair to rich soil; needs phosphorus. Its decay is of exceptional benefit to following crops. Sow early in the spring to allow time for 2 stands, 15 pounds of seed per acre. Inoculate the first time grown.

ROUGHPEA (*caley pea, singletary pea*)—winter annual legume; southern half of U.S., and the Northwest. Many soils, but best on fertile loams. Sow 30 pounds of inoculated, scarified seed in the fall. Needs phosphorus. Will volunteer.

RYE—non-legume; grown mostly in the Northeast and South. Many soil types. Sow 80 pounds in the fall. Tetra Petkus is an excellent new giant variety.

SESBANIA—legume, as far north as Washington, D.C. Prefers rich loam, but will grow on wet or droughty land, very poor or saline soils. Very rapid grower in hot weather. Broadcast or drill 25 pounds in the spring.

SOURCLOVER—winter legume; South and West. Many soils. Sow in early fall, 15 to 20 pounds of scarified, inoculated seed.

SOYBEANS—summer legume; deep South to Canada. Nearly all kinds of soil, including sour soils where other legumes fail. Will stand considerable drought. Use late-maturing varieties for best green manure results. Sow 60 to 100 pounds, spring to midsummer. Inoculate first time grown.

SUDAN GRASS—non-legume; all parts of U.S. Any except wet soils. Very rapid grower, so good for quick organic matter production. Use Tift Sudan in Central and Southeastern states to prevent foliage disease damage. Sow 20 to 25 pounds broadcast, 4 to 5 drilled, in late spring.

SWEETCLOVER—biennial legume; all parts of U.S. Just about any soil, if reasonably well supplied with lime. Will pierce tough subsoils. Especially adept at utilizing rock fertilizers, and a fine bee plant. Sow 175 pounds of scarified or 25 pounds of unscarified seed, fall to early spring. Fast-growing Hubam, annual white sweetclover, can be turned under in the fall; other varieties have their biggest roots in the spring of the second year, so turn them under then.

VELVETBEANS—annual legume; South. One of the best crops for sandy, poor soils. Produces roots 30 feet long, vines up to 50 feet long. Sow when the soil is well warmed, 100 pounds, or 25 to 30 pounds in wide rows.

VETCHES—annual and biennial legumes; varieties for all areas. Any reasonably fertile soil with ample moisture. Hairy vetch does well on sandy or sour soils and is the most winter-hardy variety. Hungarian is good for wet soils in areas having mild winters. Sown in the North in spring, elsewhere in the fall, 30 to 60 pounds, depending on the variety.

WEEDS—whenever weeds will not be stealing needed plant food and moisture, they can be used as green manures. Some produce creditable amounts of humus, as well as helping make minerals available and conserving nitrogen.

The Best Conditioners

Recent experiments reveal that after the addition of green manures and other crop residues, the soil bacteria produce materials called polysaccharides. These are the glue-like materials which stick the soil particles into aggregates so essential for a good structure.

The amount of these valuable polysaccharides, produced by decomposing green manures, can be tremendous. For example, agronomists at the University of Delaware report that decaying alfalfa and oat straw produced as much as

5,500 and 4,000 pounds per acre, respectively, of these glue-like materials only one week after they had been added to the soil!

You can't ask for much better soil-conditioning action than that.

Can Green Manure Improve Your Soil?

For a small garden, up to 50 by 50 feet, I personally don't care for green manure crops, because I like to work a small garden intensively, with succession plantings that extend far into the fall. Under this system, there is no time for growing green manure. For such a small garden, I think organic matter can be obtained in other ways, without sacrificing valuable space.

When I had a small city garden of perhaps 25 by 30 feet, I used spent brewer's hops, weeds, old hay, a very thick mulch of leaves (trucked to me free by the city) and anything else I could get in the way of organic material. This mulch of raw organic materials was applied a foot or more deep, as soon as the last crop of vegetables was removed from each row. In the spring I raked and forked this mulch, half-rotted by this time, off the garden area, to allow the ground to warm up and dry out. When the new plants were growing nicely, I replaced the mulch material around the vegetables, adding more as they grew, until all of the mulch was back on the garden. In the fall I would apply another layer of leaves, etc.

When I moved to the country, after 5 years of the practice described above, I left the most wonderful garden soil imaginable. Now that I have a garden of an acre or more, I use very little mulching, because it is too costly and time-consuming to do it well on such a large area. I can do a better job of supplying organic matter to my soil by using green manure in the fall and spring, with clean cultivation during the growing season.

My favorite green manure is winter-hardy Balboa rye. In my locality in Pennsylvania, I can plant this rye any time from early August until late September and have a very good chance of getting a stand before freezing weather sets in. In the spring, the rye grows amazingly fast and is often 3 feet high by time for spring planting. I seed the rye very thickly, and when it is two or three feet high it takes good mowing equipment to cut it.

Rotary mowers shred the rye, which makes for rapid bacterial action and decomposition.

Another green manure crop I like is hybrid Sudan, although it has some drawbacks. When plenty of moisture is available in the early fall, Sudan will make a tremendous amount of green manure in a short time. However, as Sudan is easily killed by frost, and as it grows readily only in warm weather, it must be planted fairly early. I have had Sudan over 6 feet high, and almost as thick as a good stand of rye. I mow it with a rotary mower in the same way as the rye, but sometimes it is necessary to go over the field several times to do a good job. If it is still early enough, I seed rye after the Sudan to get a spring green manure crop. If not, I leave the Sudan debris on top of the soil. In a dry fall the Sudan seed fails to germinate—or, if it does germinate, the growth is not enough to make the planting worthwhile.

If you have plenty of land, you can divide your garden into two plots, using each plot in alternate years. On the plot not being used, you can grow at least two crops of Sudan during the spring and summer, and then a fall planting of rye for a spring manure crop. Buckwheat can be used in the same manner—with successive plantings, one after the other.

There are worse green manure crops than plain weeds. I have read, and believe, that some weeds extract certain nutrients from the soil and make them available to cultivated crops. However, one word of warning on using weeds for green manure: *The weeds should not be allowed to go to seed.* Disc or mow the weeds before seeds form; then let another weed green manure crop grow from seeds already in the soil. Eventually most of the weed seeds will germinate, and you will have less trouble with them when you go back to gardening the plot. In planting one area of my lawn, I disced it every week or so, as the weeds germinated, all summer long. Finally, in August, I planted my grass seed.

This is the best piece of lawn I have, with an almost total absence of weeds, because all the weed seeds had germinated and been disced during the summer. On adjacent plots, where I simply plowed and planted grass seed in the fall, I have quite a few weeds.

—James Hare

QUESTIONS AND ANSWERS ABOUT GARBAGE

Danger of Using Raw Garbage

Q. For several years I have been putting non-greasy, raw vegetable refuse through my grinder. The liquid that drips from the grinder I dole out to seedling plants just sprouted indoors. The chopped refuse scattered over the surface of our kitchen garden quickly and inoffensively becomes part of the soil. But, it seems to me that strange, new aspects have appeared as hot weather arrived. Could these have developed from eggs hidden in vegetables shipped into Illinois from Texas, Arizona, California and Louisiana?

A. Your trouble arises probably from putting raw material on the soil. First compost the garbage. Only finished compost should be used on the land where crops are growing. Even the liquid that drips from the grinder should not be used on your seedlings. Impregnate your compost heap with it and let it decay in it.

Lemon Rinds

Q. How is one to know if *grapefruit and oranges are sprayed?* You say the skins can be used in compost, but you didn't say if it is wise to use *lemon rinds*, however I have used some.

With most everyone using poison on everything nowadays, potatoes, apples, cabbage, etc., and we who do not raise enough from our own gardens to can, should we just burn the potato peels, and leaves of cabbage, etc., and not run the risk of getting the poison in the compost?

A. Citrus fruit orchardists usually resort to poison sprays. Experience has, however, shown that lemon and orange rinds, if properly composted, do have little unfavorable after-effects.

As you say, it would be desirable to use composting material free from metallic poisons and we hope that more people like yourself will become interested in organic gardening so that the final goal of healthful food can be more widely attained.

Raw Garbage

Q. I have read a statement recently that finely cut garbage may be used as a fertilizer. What would be your opinion of this?

A. The statement that finely cut garbage may be used as a fertilizer is not complete. The finely cut garbage ought to be fermented and put through a compost heap before it is put on the land, otherwise it acts like green manure and often reduces the crop rather than improves it.

Q. We eat many grapefruit and use lemons daily. During the winter I throw the rinds on the garden, which amounts to a dozen or two a week. Can this be harmful to the garden?

A. Certainly it can. It should be composted first. The bacteria that break down the raw garbage feed on certain elements in the soil, especially nitrogen and create deficiencies. See above regarding methods of composting garbage.

Dishwater

Q. It has been recommended that farmers should take their dishwater and throw it on their compost heaps. Would that be advisable in view of the strong alkalis that are contained in soaps?

A. I do not recommend that dishwater be thrown on the compost heaps, as it often contains greasy matter which would interfere with the air supply to the roots.

Kitchen Drain

Q. Our garden is on a terraced hillside. We are building a new drain to our kitchen sink. Would it be practical to let it drain into the "compost pile" at the foot of the garden; it would be tiled to the "pile" or pit made of cement blocks.

The pit has a pipe for overflow of water that will run into a tub or barrel that can later be used for watering plants that need it. Is this practical? Would there be enough fertility in this water to pay to carry it to the plants?

A. Unfortunately kitchen drains carry a goodly amount of fatty substances which are not easily broken down by soil bacteria and thus prevent the formation of good compost. There is another consideration: Since you could not adequately control the influx of water from the sink, the chances are that your pit would become waterlogged; inevitably, the nutritive substances, especially Nitrogen, would be leached out and lost to your garden.

As to watering plants with kitchen water, the fat particles are apt to cause trouble sooner or later, since the plants breathe through their many pores, called stomata or mouths, and these would be clogged by fatty water.

Flies in Garbage

Q. I would like to save my kitchen garbage for a compost heap but, even in a tightly covered can, it becomes a breeding place for flies. How can I prevent flies from infesting the garbage?

A. Flies may be excluded from a garbage can if it has a really close-fitting lid or cover. Or the garbage can be put in the compost heap daily and covered with enough soil to exclude the flies. Try placing a sheet of newspaper over the top of the can before the lid is replaced on the garbage can.

Fruit Flies Can Help

Q. I have noticed some small, hair-like organisms in my garbage heap. How can I get rid of them?

A. The small hair-like organisms which occurred in the garbage are doubtless the larvae of certain insects, possibly the larvae of the fruit fly. These organisms, whatever they are, play an important part in breaking down garbage.

Garbage Odors

Q. Is there any safe way to eliminate the odor in a kitchen garbage container?

A. It is recommended that about two inches of rich loam be sprinkled in the bottom of the container. Then use the earth on top of the garbage that you empty into the pit. One reader observed garbage composts in a fraction of the time required before she started mixing it with a light layer of bacteria-rich earth. This practice is especially helpful when one builds a pile gradually.

Garbage Analysis

Q. What method of composting garbage will cause the least destruction of nutrients?

A. The two-foot method is about the best. Use a pit or build a heap that reaches no higher than two-feet, in which daily garbage is layered with earth. Then there will be practically no heat generated, which ordinarily burns up the organic matter to a fraction of the original amount. Allow for drainage, either by collecting it in the top few inches of soil and applying in your garden eventually, or by using a slanted cement bottom with a channel to catch the liquid. If you are not bothered particularly by animal pests, try running the liquid into a small pot, which can be conveniently used to water the garden.

You can make the fastest compost if you add earthworms to the decaying garbage. The organic matter will be sweet-smelling compost in no time.

Q. Can you tell me what the chemical analysis of garbage is?

A. Garbage, such as accumulates from restaurants, dormitories, army posts, etc. is (on an oven-dry basis) composed of about 25 per cent protein, 28 per cent fat, two per cent fiber, and six per cent ash, as experiments at the University of Hawaii showed.

A standard analysis of New York City garbage rubbish showed the composition of 3.4-3.7 nitrogen, .1-1.47 phosphoric acid, and 2.25-4.25 potash. Its relatively high nitrogen content produces quick decay.

Never Use Lye

Q. An acquaintance told me recently that he composts all his garbage using lye. He digs a hole, quite deep and throws into it all his garbage, covering it with earth. The earth in thin layers keeps the flies and rats away to some extent but he uses in addition a can of lye, dissolves it in a pail or two of hot water and pours the liquid over the garbage. He must use a fiber pail because, as he says, the lye would dissolve an iron pail. It completely dissolves all the tin cans in the garbage, leaving no trace of them. Won't the lye also kill soil bacteria and fungi that are so necessary for decay?

A. Whatever chemicals are used to drive away insect pests will almost always drive away valuable organisms in the soil.

Garbage Failure

Q. Last winter I took a small barrel and knocked both top and bottom out of it and during the winter months put into it our table garbage, occasionally sprinkling a little lime and bone meal over it. I was very much disappointed this spring to find that it had not decayed but was sour and foul smelling.

A. There are many things wrong here. If you had sprinkled rich topsoil in as you layered it, putting the soil in every two or three inches—about 1/4 inch thick, it would have controlled the odor much better. If you have finished compost that you can use instead of topsoil the results will be even better.

There is an enzactivator which is a Japanese activator based on enzyme action, for the decomposition of organic matter. An enzyme is not a bacterium. It is created by the living cell of a plant or an animal, including bacteria. Enzyme substances are capable of producing chemical changes in other substances, by catalytic action, that is without itself being changed or consumed. Eventually, through wear and tear, the enzyme disintegrates and must be replaced.

Enzymes play a prominent role in the decomposition of carbon compounds which are the most prevalent constituents of organic matter. The object therefore is to mix the prepared enzyme material with organic matter and manure so as not only to accelerate their decomposition, but to prevent the growth of harmful bacteria and the escape of ammonia and at the same time stop the disagreeable odors normally present during decomposition.

Where garbage is put in barrels in the basement of homes, if POWDERED charcoal would be sprinkled every two or three inches, it would absorb the odors.

Kitchen Wastes

Q. As a new subscriber I may have missed out on some advice you have concerning the value of kitchen wastes. What can I use of the things left over from the kitchen?

A. All kitchen wastes can be used with the exception of soapy water and fat. Soapy water contains dangerous chemicals, especially soda, which is not good for most plants. Only beets like it well. Fat, besides being an attraction to ants, does not break down very well and had best be used elsewhere. That leaves us with garbage of all kinds, scraps, bones, etc.

The other kitchen wastes had best be thrown into a garbage pit. Pit or pile should be occasionally watered and covered with earth to speed up decay and to prevent fly trouble and odors. Then when you have enough green matter collected, you can use this half-decomposed garbage as an excellent activator. Depending on how much animal matter may be contained, your garbage will be more or less rich in nitrogen. Hotel garbage, leftovers from mess halls and army camps are noticeably richer in meat scraps than private kitchen wastes and are therefore rich enough in nitrogen to supply all the animal matter needed to stimulate bacterial growth. But lean private kitchen wastes are low in nitrogen and you will therefore have to use a few bags of poultry manure or similar sources of animal matter to get a good compost heap.

Q. How shall I handle garbage in making compost?

A. Since garbage contains green matter mixed with animal matter, it should be handled a little differently. If it contains enough meat and fish residual value, it can be used practically without manure, but a little should be used because manure contains a lot of bacteria which aid the processes of composting. In such case use six inches of green matter and four inches of garbage, then about a half inch of manure. Then sprinkle lightly with earth and lime.

If the garbage does not contain animal matter, use six inches of garbage as the green matter, then two inches of manure. Then a sprinkling of earth and lime.

Suburban Composter

Q. Could I make a compost heap in a new galvanized garbage can? I have only a small backyard and not much room.

A. A garbage can could be made into an ideal composter. Punch holes in the bottom and in the sides as well as the lid. Place about 4 inches of fertile soil in the bottom of the can. Then layer your materials in the same way as for a large heap in the garden. Such a composter is not objectionable even in the best residential parts of the city. If desired, plants can be grown and trained over the can so that it becomes converted into a mound of green leaves.

Composting: "Everything Tastes So Much Better"

Chapter 9

Scraggly transplants began forming Herculean leaves and stalks. Late-started sweet corn, with aluminum foil set around the plants to reflect needed light and to repel aphids, also responded with a burst of growth. So did several more of the thoughtfully fragrant ornamentals—tuberoses, lily-of-the-valley, nicotiana and a BLAZE rose hedge.

In the vegetable patch chard, spinach and tomatoes felt their oats while a row of strawberries, a red currant bush, a newly-planted Carpathian walnut and a dwarf peach and apricot espaliered against the house have all grown appreciatively. More than that, "Everything tastes so much better," emphasizes Mrs. R.

What caused all this to happen?

Compost.

Here's how Paul and Mildred Robinson got their amazing results. Paul Robinson built a compost box. Not an ordinary one, mind you—his was of solid wood construction, befitting his training as a mechanical engineer. Set in a corner of the small yard, the snug 6-by-4-by-3-foot bin was made both airtight and waterproof, then given an aesthetic dark-green coat of paint.

Everything was ready, including the Robinsons, who sailed forth with high hopes. The only hint of impending turbulence was one neighbor adjacent to the corner where the new bin had been erected. When he dourly commented, "You're not going to keep that thing there, are you?" they quickly assured him there'd be no problem. And there wasn't, *ultimately*—but not until a few hectic days had elapsed and a few basic ideas about composting had a chance to take root.

Ingredients for a Heaping "Supercompost"

By the start of April the Robinson heap got under way —somewhat. Mildred had worked out what she considered a "supercompost" recipe, one which piled in just about everything *including* the kitchen sink's wastes. A graduate biologist, Mrs. R. applied scientific zeal to her search for organic materials to enrich the humus that new bin was to yield. Besides their own table scraps and coffee grounds, she assembled all the leaves, weeds and grass clippings around the place, spreading these in 2- to 4-inch layers inside the box, along with some sawdust, ground limestone, a little "miserable clay" soil, and small amounts of bone meal, horse manure (from a nearby riding stable), dried cow manure, and water.

Two weeks went by; and nothing much seemed to be happening. The pile hadn't heated, but just appeared to be reclining in total inactivity. One difficulty (the Robinsons learned later) was that a lot of the plant wastes—stems, stalks, roots, etc.—were coarse or woody and rather big for the proportion of the heap. (Shredding these materials, even a little by running a power mower over them, would have speeded up the decomposition process by exposing more pieces and sides to the natural decay bacteria and fungi present.) Also, they might have been keeping the pile too moist, perhaps soggy, in the compact, well-sealed bin.

Anyway, once the two-week period passed, the mail brought the booklet on composting. "We decided we didn't have enough nitrogen," recalls Mildred, "and, oh boy, we went wild adding it." What they added was an imaginative assortment of organic materials rich in nitrogen: cottonseed meal, blood meal, aged sheep manure, more bone meal, more horse-stable manure, and even the family's hair clippings ("We trim our own") which they read were a valuable source.

Along with these additions, the Robinson recipe took on such ingredients as peat, granite dust, wood ashes, powdered phosphate rock, old flowers from the spring border, wool shavings, more dolomitic lime and coffee grounds, plus some leftover wheat germ that had turned rancid. On top of all that, Mrs. R. phoned around to the area's breweries to turn up one that offered spent hops for the taking. So take them they did, supplementing their already well-diversified heap with a generous layer of the sopping hops.

The Slumbering "Volcano" Erupts

Another 72 hours slipped by quietly. The Robinsons bought a batch of earthworms from a local bait shop, and introduced them to the slumbering binful, which they had started turning.

Within a day's time, the "mountain" erupted. "We got the heat then, all right," exclaimed Mildred, "billowing, smoking heat, which sure impressed us, since we'd been

Coffee grounds are a valuable addition to the compost pile for their supply of nitrogen.

unsuccessful in getting any at all!" That's also when the "scent problem" began—and that made quite an impression too. "All we could think of were the neighbors," groaned Mrs. R., "and all they got was a spell of this heavy odor." The whole thing was maddening, she continued.

"We started sneaking out at night to turn the heap, figuring the more oxygen we let in, the faster it would decompose." True, but unfortunately this also added to the aroma now permeating the neighborhood—at least until the breakdown of the coarse, wet materials caught up with the usually odor-free compost process. It became a task-plus to slink out late at night to try turning the heavy, entangled mass, made heavier by coarseness and excess moisture.

Just about the time the Robinsons verged on becoming discouraged—"and sort of muscle-sore at so much nocturnal turning"—their reluctant compost heap mellowed. Despite the disadvantage of large amounts of chunky, hard-to-start-decaying materials, the pile responded to the nitrogen- and bacteria-rich ingredients so liberally incorporated. With a little shredding, and possibly a bit less compacting and extra watering, even this ungainly gathering of compostable matter could have been transformed to fine humus —without the nose-holding neighbors calling "foul"!

So, after the Robinsons gloomily had visions of needing to have their entire "super pile" hauled away (just "wasted waste," as Mrs. R. remembers thinking of it)—things suddenly straightened out. The air cleared, so to speak, and their multi-enriched compost reached a state just right for the waiting garden.

Once the "barn fire" of the heap slowed down, and the overnight fragrant bloom of their peonies helped out with the subsiding scent the innovating couple found their hard-earned compost dark and mellow. They eagerly started to try it out in different places.

You know the results.

"Take a Lesson from Us"

One vital point the Robinsons stress is not getting dismayed. "If anyone ever feels discouraged, I hope they'll take a lesson from us," urges Mildred. "By now I feel almost well-qualified to offer some advice—and encouragement. Whatever you do, don't give up on compost. Find as many materials and organic by-products as you can in your locality. Watch the bulky materials and layer proportions. Try to break up the coarse wastes. We've sent for plans to convert an old mower into a shredder.

"Organic gardening's been such a boon to us, we hope to get more people interested—even by way of our off-beat composting experience!"

Another Successful Composting Program

Warner Bower's first garden project in his new Long Island homestead was the contruction of a good-sized, working compost pit. He chose six-foot lengths of 1-by-10 lumber, which gave him about 144 cubic feet of space, piling the 10-inch wood five boards high. Each section is removable as the level of the compost pile rises and falls with the advance of each season.

The system obviously has its disadvantages—the finished compost is always on the bottom, and a strain on the back when you want to get at it. So he added a second pit, this time of galvanized iron, which loads from the top, and discharges finished compost from the bottom when its redwood bars are shaken, just like emptying ashes from a coal stove. He uses the two types of pit somewhat differently, and each serves its own purpose. Since the smaller pit has a tight lid, he refers to the pits as the open and the closed, to distinguish them.

Garbage Is Composted

All wet garbage and organic kitchen matter, except meat scraps and fat which tend to putrefy and have a disagreeable odor, are composted. Even the meat scraps and fat are useful, however, because most of this material goes into our six fat-feeders which serve hordes of bird friends, who gobble it all up all year long before decomposition can cause a nuisance.

Paper, such as coffee filters, goes into the compost, as do the old paper sacks from our bagged fruit. He hasn't composted newsprint although there is no reason not to, if it is shredded. Warner uses newspaper in eight layers under mulch around trees and shrubs, where a grass- and weed-free area is desired.

Since the bulk of his compost is from leaves, grass and vegetable debris, he lacks the bacterial elements which cause the pile to heat up and decompose rapidly. But because he uses the compost only in spring and fall, his slower pace of production is no problem. He has tried the commercial bacteria additives and found them valueless. Earthworms abound in his compost and are added whenever he can catch one in the garden. Mr. Bower adds a sprinkling of bone meal or lime to the pits at intervals.

Disagreeable odor and flies never have been noticeable, and his compost usually smells strongly of citrus fom the many grapefruit and orange peels which he cuts up in the garbage. This pleasant odor is noticeable even when compost is spread in the yard. Warner finds the backbreaking work of turning the pile objectionable, and does not do it. Each six months, when he wants compost, only the thin top layer need be forked aside, and the remainder is usable.

In the big open pit Warner depends largely on whole leaves, ground leaves, grass cuttings and weeds. Twigs, branches and garden vines are added only after they have been ground, because otherwise they decompose so slowly that sieving the compost is necessary. With this grinding procedure, use of chicken-wire sieve usually is not necessary. Originally, he put all wet garbage into the big open pit, but soon discovered that he got little profit from it. Blue jays took all the eggshells and many fruit or melon rinds; other birds took their share, and it attracted a steady nightly stream of opossums and raccoons with early-morning visits from squirrels and chipmunks. Now he puts garbage into the open pit only if there are leaves or dirt for a covering layer. Enough moisture for the pile is furnished from rain and the frequent garden sprinkling.

The closed pit measures 25-by-25-by-10 inches, holds

about 15 cubic feet, and has been carefully set up so that all leaves are ground, and no coarse material is used. He fills the closed pit with alternate layers of wet garbage and ground leaves, with occasional sprinklings of lime and bone meal. Since the closed pit is not accessible to rainwater—the open pit is—Warner usually adds an equal volume of water each time garbage is added, which keeps the pile sufficiently wet. When shaken out, this compost does not require sieving.

"Our use of compost runs to about 70 bushels twice a year—spring and fall," he reports, "but we use none in the garden where ground leaves, salt marsh hay and other mulch material take care of the soil. In the fall, we use the compost in October just before we empty the pit for ground leaves. In the spring, need for an empty pit before we grind the winter accumulation of leaves demands we use our compost in early April. The spreading program is the same, spring and fall, and is as follows:

"1. *Orchard*—Each of the 22 fruit trees gets a bushel of compost within a radius of three feet around the trunk in the fall, covered later by salt marsh hay, but left uncovered after the spring application. Some of the heavier-bearing trees are given two bushels of compost. In addition to being great fertilizer, this material holds water very well. We put the compost mulch right up to the trunk and, although there are field mice around, they have not done any damage to the trees.

"2. *Flower Beds*—Seven large flower beds get a fall sprinkling of compost, scattered but not in a solid layer. Since we have completely converted to perennial flowers, it is necessary to plan for their ability to push through a not-too-heavy layer. The water-holding capacity of the compost is readily apparent in comparison with non-composted areas which are dry a day after heavy sprinkling.

"3. *Roses*—The 35 rose bushes get a half bushel of compost each, twice a year, covered by ground leaves. The application of ground leaves is carried all along the fence, in a two-foot-wide, four-inch-deep layer which holds moisture, and kills grass and weeds, making trimming or weeding rarely necessary.

"4. *Trees*—We throw a bushel of compost under each of the 14 blue spruces. Around most of our large oaks and hickories, we have plantings of day lilies, cannas and various kinds of hosts to make mowing and trimming easier. Over these plantings we can spread compost because these strong perennials will push up. But around some trees we have a mulched area of pine bark planted with pachysandra, which does not die back, and which therefore cannot be composted easily.

"5. *Shrubs and Vines*—The various specimen shrubs and vines lend themselves well to application of compost in proportion to their size. Although our compost is neutral in reaction, we add a little lime around such acid-haters as the lilacs.

"All in all, our need for compost runs to about 70 bushels twice a year. By composting all available vegetable material and almost every scrap of organic kitchen waste, we meet our requirements and do our share in helping to avoid pollution of the environment."

Cauliflower + Compost = Nine-Inch Heads

Cauliflower tests the skill of any gardener but particularly those who must garden in the hot, dry Southwestern states. Still, my success with Brussels sprouts last fall—which are no picnic to grow in this climate either—made me want to try my hand with this delectable vegetable.

Around the latter part of July I dusted off my little greenhouses, filled them with a mixture of compost, sand and peat moss, and on the first day of August planted my seeds.

To conserve moisture I kept the covers over my greenhouses, and in a few days the husky little plants were peeping through, their leaves a good, healthy, deep-green color, sturdy and compact. Now I took off the covers, and set the trays where they would have the morning sun. They grew out well, were soon showing the third leaf, but were still quite small.

With my husband's help we cleared the garden area

where the early corn had been planted, plowed it up, and pulverized it with the rotary tiller into a fine seedbed.

I then marked off several rows and set the plants about 15 inches apart. The rows themselves were 24 inches apart. Since cauliflower grows to a large plant I wanted to give it plenty of room.

The plants continued to grow well and I watered them about once a week, thoroughly soaking the soil. Mulching them with grass clippings helped to conserve the moisture and keep their roots cool. The first of October we received a good general rain and then I felt I could rest a little and let them proceed to grow on their own.

About this time Carl decided to start taking down the compost heap where our cantaloupes had grown earlier and he began putting this on quite heavily around my cauliflower plants. He had built this up to a depth of several inches by October 16, the day we left for a two-week vacation in Utah.

When we returned, late in the afternoon of October 29, I ran out to the garden even before I went into the house to see what my cauliflowers looked like. I had the surprise of my life. Almost every plant in the three rows was wearing a snow white crown which ranged in size from six inches to one enormous head nine inches in diameter.

This spectacular growth was almost unbelievable and I am sure the heavy layers of compost were responsible for this minor miracle which occurred during our absence. No doubt the cool, moist weather had also contributed but it had been the nutrients in the soil which had done the most to promote the growth of my cauliflower crop.

I had all I could possibly want for table use and it was a simple matter to process and freeze the surplus heads to be enjoyed all winter long. There was even enough so that I could reserve several heads for pickling with my late fall cucumber crop, giving my pickles more eye appeal and making them extra delicious.

Yes, even a difficult crop like cauliflower, which I could not possibly grow with success in our warm climate in the spring—can easily be achieved by correct timing. And since this vegetable sells for 40 to 50 cents a pound in our local supermarkets it appeals greatly to my thrifty nature to have all I can eat—*free.*

With *plenty* of *compost*—it's *that* easy!

—Louise Riotte

The Fourteen-Day Method

Perhaps the simplest, most well-known, and most efficient method of composting is the fourteen-day method.

You can make 18 *tons* of crumbly compost in six months, in a space *8 x 4 feet!* This is enough compost to put an inch layer—an ample annual dosage—over 6,000 square feet of garden area, or a space 75 by 80 feet!

How is it done? Easily—using a shredder or rotary mower with the 14-day composting method, you begin in the spring, making a 4-foot high heap in a space 8 x 4 feet. After 3 weeks (we have allowed an extra week for possible heating), remove the finished compost—110 cubic feet, weighing about 2 tons—and begin the next pile. By mid fall—6 months later—your ninth heap should be finished, giving you a 6-month total of 18 tons.

The 14-day method has passed many trials with flying colors. Initially devised by an agricultural team at the University of California, the method was tried first at the Organic Experimental Farm in 1954, and since that time many readers have reported success with it. Here is the basic formula and procedure. This is not absolute, of course, and it must be altered to suit your individual needs and supply of material.

Day-to-Day to Two-Week Compost

First day. Your basic material can be one of a number: leaves, spoiled hay, weeds, or grass clippings. To these should be added one of the nitrogenous materials (manure is best) and any other material available. In the editorial experiments, equal parts of leaves, grass clippings, manure, with a liberal sprinkling of natural rock powders, were found to work very well. Remember, too, that leaves tend to mat down, slowing the composting process; they should be mixed with other material. Shred everything including the manure, with a compost shredder or a mower; mix materials together and place in a mounded heap. If your materials are low in nitrogen, be sure to add a sprinkling of dried blood, cottonseed meal, or other nitrogen supplement on each series of layers.

Second and third days. By now, the heap should have begun to heat up. If not, add more nitrogen. Bury a thermometer in the heap to check temperature. Keep the heap moist, but not soggy.

Fourth day. Turn the heap, check temperature, and keep moist.

Seventh day. Turn it again, check temperature, and keep moist.

Tenth day. Turn it once more. The heap should now begin to cool off, indicating that it is nearly finished.

Fourteenth day. The compost is ready for use. It will not look like fine humus, but the straw, clippings, and other materials will have been broken down into a rich, dark, crumbly substance. You may want to allow the heap to decay further, but at this stage it is perfectly good for garden use.

What's the Best Bin?

The best compost bin is the one that turns your homestead's organic wastes into soil fertility quickly, cheaply, and easily.

The materials don't matter, but the results do. For years, organic gardeners have been making compost bins with bits of old wire, scrap wood, discarded lumber, old cement blocks, and battered, secondhand steel drums. They've been using whatever they could lay their hands on (free donations are always appreciated!) to make compost bins that fitted in with their gardening programs. Here's how they've been doing it for the last ten or more years.

A few years ago, Lyman Wood built a wire and wood bin using scrap 2-inch-square lumber, which he covered with ½-inch chicken-wire mesh. Made of 2 L-shaped sections held together with screen-door hooks, the cage provided him with 18 to 24 cubic feet of finished compost in 14 days, which was par for the composting course then.

Composter Wood reported that the pile heated up to 140 to 160 degrees in 2 days, and could be turned in 4. He damped down each layer—leaves, grass, garbage, and manure—as he added it, and counted on the well ventilated cage to encourage complete bacterial action.

Turning was extremely easy. He unhooked the sides, separated each of the L-shaped sections, and then reassembled them next to the square sided heap. "You will be pleasantly surprised at how neatly and firmly the heap stands," he wrote, adding: "It is now a simple and satisfying task, using a fork to peel the layers off the pile and toss them in the now-empty cage." During the turning operation, he kept a hose handy to wet down the heap as the material was transferred.

"I have used this cage many times," he concludes, "for 14-day rapid composting, and each time it is a satisfying successful venture to harvest the cube of dark, moist crumbly humus."

The same idea—a portable compost bin that can be lifted, leaving a pile ready for turning—was used about six years ago by the Peter Seymour Company of Hopkins, Minnesota. The "Cake-Maker" was made of wood and metal framing, and was light enough to be lifted off the pile, leaving a "cake" of compost, and ready to be filled again.

Bales of Hay Make a Winter Compost Bin

Alden Stahr once made an all-winter compost bin out of old bales of hay stacked around secondhand storm doors and windows which he put up in the garden. Composter Stahr mixed in garbage and manure to help heat up the pile, while the glass lid—slanted to the south to pick up the long, low, midwinter rays of the sun—also kept out scavenging cats and dogs.

Although he recorded a 50-degree difference in temperatures between the inside of the bin and the outdoors one cold January morning, he achieved an extra supply of compost, which he had ready for early-spring use, "thanks to the billions of happy bacteria hard at work."

The New Zealand bin—air from all sides

The classic among compost bins is the wooden New Zealand box which was originally designed by the Auckland Humic Club to admit as much air as possible from all sides. Many variations exist, so don't hesitate to change your design to fit your material and budget.

The important factor is air circulation and ventilation from all sides, so be sure to leave 1-inch spaces between your slats or boards. It's a good idea to start with a rugged frame work—two-by-fours are excellent—and then nail a lattice of boards over it. The top and bottom are left open, although some composters prefer to cover the top of the pile in rainy weather to prevent leaching. One or two good coats of linseed oil should be allowed to soak into the wood to make it weather and rot resistant.

Steel Drum Composters for the Suburbanite

West Coast gardener John Meeker reported several years ago how he solved the twin problems of running a compost pile in a congested suburban area without offending his neighbors, while getting enough compost to run his garden. He circulated air into the heap, using a steel drum, and had it raised 6 inches off the ground by setting it up on a circular metal frame with legs.

Meeker reported that the construction has several advantages over the piles and pits that he used before. *The air can circulate up from the bottom of the barrel.* The six-inch space allows easy removal of the compost. The moisture content of the compost can be carefully regulated by covering the barrel with a lid . . . leaching of the compost is perfectly controlled. *By simply covering and uncovering the top, one can regulate the amount of air introduced into the mass.*

There is always a bushel or two of compost ready, Meeker noted, even with so small a composter "once the cycle has begun." This includes such seasonal bonuses as summer grass clippings, autumn leaves, and crop residues which are "ready to enrich the garden by the time one gets ready for spring planting."

As for the neighbors "who have gladly shared" his bumper harvests while "turning up their noses at my deposits of leaves, cuttings, manure and—worse—garbage," Meeker reported the solution to the odor problem.

"When I have a large amount of lettuce leaves, beet tops, grass cuttings, or kitchen refuse, I *whiten the top of the dampened pile with a sprinkling of ground limestone,* and over that I add a thick layer of dried steer manure. *"The limestone helps to decrease the smell and lessen acidity of the green refuse and garbage."*

A more complicated application of the steel-drum composter calls for nesting one drum on the bottom third of a slightly larger container, and installing a metal lattice grate between them to hold the pile up so air can get at it. Built by Ralph Poe of Canton, Illinois, the drum composter also featured a hollow, vertical, 3-inch-wide pipe with ¼-inch perforations that was thrust down into the heap's center and left there for additional ventilation.

Raising Pile off Ground Eases Circulation of Air

You don't have to build a revolving drum to get air into the center of your compost heap. Just raise it off the ground—10 inches is fine—by building a substantial open lattice support right into your bin.

You can also make a wooden base for your compost pile, reinforcing it with ½-inch netting to get it the recommended 10 inches off the ground.

Researchers at Phoenix, Arizona, found that a ton of rapidly decomposing compost uses up 18,000 to 20,000 cubic feet of air daily, and that turning of the pile doesn't always get the job done. But "forced air" composting—raising the pile off the ground—stimulates uniform decomposition of the entire pile, not just the top 12 inches.

The "open-hearth-bottom" bin is made of wood with a

sturdy grid of 1-inch piping holding the compostable mass 1 foot above the ground. It's 4 feet square, made of sturdy secondhand lumber. A user reports never having to turn the pile, and advises that you don't have to worry about "mass and bulk as long as you have the open-hearth-bottom, and always have plenty of good compost."

The Compost Bin Will Pay Its Way

A compost bin will soon pay for itself in reduced garbage disposal expenses. But there is more to composting than that. A compost bin pays for itself in a more productive and beautiful garden, and tastier, more healthful food for the entire family.

The bin should be inexpensive to construct, and it shouldn't take more than three hours to make. Use whatever materials are cheap and abundant in your area, and don't be afraid to accept handouts. Put the bin where it will get both sun and air, and is handy both to the garden, where the compost will go, and the driveway, where so much of its material will come in.

Finally, every homestead—the little ones as well as those in crowded towns—should have its own compost-making bin in this time of widespread and mass pollution of the environment. Remember, composting is the only safe way to handle the family's organic wastes and leftovers.

Better start building that bin—now.

—Maurice Franz

Organic Fertilizers: Where to Find Them

Chapter 10

Increased activity in the large-scale manufacture and distribution of complete organic soil conditioners and fertilizers is helping you garden organically.

In the past, the point was made all too frequently that not enough organic materials were available to warrant a mulching and composting operation.

Today, thanks to recent developments, the supply of organic materials is greater than is generally realized. There is a marked trend to the fullest-possible utilization of organic waste products on a regional basis, coupled with their mass distribution to an ever-widening market. The 3 major sources of organic materials now include:

1—The gardener's own home-grounds;

2—Materials obtainable locally;

3—Products shipped in from other areas.

First Source of Supply

The experienced gardener knows that his homegrounds should be a prime source of soil nutriments and conditioners. Kitchen garbage or refuse alone can be the nucleus of a continuous composting program when it is combined with crop residues, weeds, leaves and cuttings plus the leftovers from the annuals. These soil replenishers fortunately occur in consecutive, seasonal order, and the wise homesteader arranges his composting and mulching program to take full advantage of the cycle.

To the resources of the homestead may be added many organic materials which are available locally. Much is to be had free for the asking or taking. Road crews are happy to deposit whole truckloads of wood chips because it saves them the trouble of driving miles to the dump. Leaves in wholesale quantities are to be had, not only in the fall, but any time of the year, if you'll drive to where they're windrowed. Saturday morning is the best time to clear sawdust out of the local mill. Spent hops may be had for the asking at the brewery, but it's a good idea to do your hauling late in August when the pile is quite dry.

Regional Organic Resources Are Now Being Exploited

The amount and diversity of organic materials, once treated as waste but now processed and made available, is considerable.

In New England, wool shreds from local mills are offered as a low-cost source of nitrogen. In Texas, cotton gin trash is composted instead of burned. Ground oyster shells are used in Baltimore, filter mud in Louisiana; volcanic minerals are mined and processed in California. Along the seacoasts, kelp and fish wastes are being converted into fertilizers and soil conditioners. In Georgia, where granite rock and poultry manure are abundantly available, a low-cost, complete soil conditioner, enriched, activated Hybro-Tite, is being processed, refined locally and bagged for national distribution. The experience of its makers, Blenders Incorporated is representative of what is taking place all over the country in the treatment of organic waste products.

Granite juts out of the earth in Georgia and has been used for years in construction and paving because of its low-cost availability. But about 30 years ago Charles L. Davidson, president of Stone Mountain Grit Co., became aware of a paradox. His company was selling more granite grits—a waste product—to the poultry trade than it was selling structural stone.

On investigation, the lowly waste product was found to rate extremely high as a poultry feed. It supplied the right amount of mineral needs to laying hens, which were moreover attracted to it by the sparkle of the small specks of potash-bearing mica.

Subsequent laboratory analysis revealed that the granite that goes into Hybro-Tite is a hybird granite gneiss which contains all 22 trace minerals and assays 5.2 per cent potash. Hybrid granite is an amalgam of many different rocks formed millions of years ago under terrific heat and pressure, bubbling up out of the earth like lava. Seen in the Davidson quarries near Lithonia, the granite is streaked with different-colored veins, ranging from greenish-grey to rosy pink.

Until recently, poultry droppings were considered a glut, an economic liability, in Georgia where 50,000-hen poultry farms are common. The Davidsons, who supply just such farms with granite grit, were familiar with the situation. One day the question seemed to ask itself and to supply the answer simultaneously: "Why not combine the granite with the droppings?"

Today chicken droppings are hauled by the truckload in from local poultry farms and mixed with granite grit in 10- to 20-ton batches. Activated with a bacterial starter, they are then spread in windrows 5 feet high by 5 feet wide and allowed to heat up to 160 degrees Fahrenheit.

Composition of the finished product, enriched, activated Hybro-Tite, is no secret. The components of one ton of the finished product include:

CHICKEN COMPOST	600 lbs.
GRANITE MEAL	1000 lbs.
COLLOIDAL PHOSPHATE	100 lbs.
GROUND MEAT SCRAPS	100 lbs.
HUMUS	200 lbs.

Composting is a strictly controlled process. The windrows are turned and aerated at 160 degrees to check further temperature rise and the moisture content of the mass is carefully checked. Joseph Francis, president of Blenders, Inc., was careful during a tour of inspection to point out long mycelial strands of thermophilic mold in the heated piles and the subsequent appearance of "fire fangs" or silver flecks of actinomycetes in the heaps as they cool off, signs of satisfactory decomposition.

The Davidsons believe they have achieved a genuine organic breakthrough by combining two waste products in an area where they are plentiful to create a complete soil conditioner. They are certain their results can be duplicated elsewhere and plan to hold seminars in which their process of combining rock minerals with manure and composting them is demonstrated, step by step.

Another large-scale manufacturer and distributor of organic soil conditioners, the Natural Development Company, combines 18 different organic substances at its home plant in Perry Hall, Maryland. Fertrell, their product, includes blood meal, castor pomace, greensand, tobacco dust, oyster meal, chicken manure and various sea products.

Fertrell, like Hybro-Tite, is a soil conditioner which provides humus and has a rich supply of trace minerals. It is packaged and offered in 4 blends, ranging from ratings of 1-1-1 to 4-3-4.

Municipal Sludge Another Source

As we have seen, large-scale processing of municipal sludge makes available still another organic material that frequently may be had free for the taking. The nitrogen content is from one to six per cent. Where it is not free, sludge costs from $4 to $70 per ton, in many cases delivered. A 100-pound bag sells for about $1.15.

Potential Much Greater than Believed

The potential of animal manures alone as a source of true organic fertilizers and soil conditioners is actually greater than is commonly realized. The annual value of poultry manure produced in Connecticut alone in 1960 was over $1,500,000. The corresponding figure for all New England in the same period was well over $6,000,000.

"A multi-million-dollar manure pile that could go a long way toward keeping Ohio green." That's how G. E. Mountney of the Ohio Experiment Station described the one million tons of poultry manure produced by Ohio's laying and broiler flocks.

Nevertheless, the attitude persists that animal manures are scarcer today than they were 35 years ago when the horse was still the prime source of power on the farm. Livestock is up, over 40 million head, to offset the loss in sheep and horses of about 36 million. Meanwhile the poultry population has risen phenomenally—there were about 370 million chickens in 1963. It has been estimated that since 100 hens produce slightly more than 7 tons of manure a year, about 26 million tons of nitrogen-rich manure are produced annually in this country by its poultry.

But chickens are not the only source of organic animal fertilizers. Last year 300,000 tons of steer manure were bought in California alone—the result of an all-out drive to process and package for mass distribution a waste product that had formerly been an economic glut. Quality of the end-product has been criticized since as not uniform, the salt content was reportedly high, and prices fluctuated widely. But a breakthrough had been achieved on a large scale as the cattlemen showed what could be done by appealing to the mass organic market.

In the future, we expect to see more such marketing of low-cost packaged animal manures. Lagoon disposal, at first

APPLYING ORGANIC FERTILIZERS

Here are some practical ways recommended by experienced gardeners:

1. Spreading the fertilizers in the seedbed *before planting*
2. Sowing them along the seed row *during planting*
3. Setting fertilizers in or around the hill *before or at planting time*
4. Applying fertilizers along the plant row *during the growing season.*

Rate Per Acre, in Pounds

	250	500	750	1000	1500	2000
12″	9 oz.	1 lb.	1½ lb.	2¼ lb.	3½ lb.	4½ lb.
15″	12 oz.	1¼ lb.	2 lb.	2½ lb.	4 lb.	5 lb.
18″	14 oz.	1½ lb.	2½ lb.	3 lb.	5 lb.	6½ lb.
24″	1 lb.	2 lb.	3 lb.	4½ lb.	6½ lb.	9 lb.
30″	1¼ lb.	2½ lb.	3¾ lb.	5¾ lb.	8½ lb.	11 lb.
36″	1½ lb.	3½ lb.	4½ lb.	7 lb.	10½ lb.	13½ lb.

Amount of Fertilizer Per Row for Various Rates of Application Approximate amounts per 100 feet, for rows different distances apart touted as an ideal solution to the manure disposal problem, has run into several snags. There just isn't enough room on most farms to build ponds large enough to handle animal wastes safely. In addition, vast quantities of water are needed to dilute the manure to the right proportion for conversion.

—Maurice Franz

Nowadays Everybody Stocks Organic Materials

We went shopping for organic fertilizers and mulches one wintry day so we could tell you what's available right in your own backyard even on one of the most unlikely days of the year.

We wanted to find out what you can buy, how far you may have to go to get it, and just how hard it might be to get some pretty scarce items—seakelp and bagasse, for example. This is what we found out.

Nowadays everybody stocks organic mulches, fertilizers and soil conditioners right along with the chemical stuff. All you've got to do is ask and, not only do they know what you're talking about, they've got it, and at a price that's competitively right.

These are some of the places we deliberately passed up although we knew we could get all the free materials we could use:

1. The feed mill that gives you all the free ground corncobs you can haul away
2. The city dump that lets you take all the leaves you want
3. The brewery that keeps spent hops in the yard—free for the taking
4. The lumberyard with a sawdust shed that's handy for backing into
5. The city sewage treatment plant that fills your truck with sludge for $1 a load

We passed these places on the road—didn't have to detour—but didn't stop because we were on an "organic shopping spree." Regarding that argument, "It's hard to get organic materials," we can only stress that we could have made 13 calls within our self-imposed 25-mile radius.

Plenty of Organics at Garden Supply Center

Our first call at a real "old country" garden supply store revealed a good variety of meals—blood, bone, and cottonseed—plus cocoa bean shells, shredded pine bark, and a soil amendment and mulch called TURFACE which was new to us. Technically described as a "montmorillonite clay," it seems to be Mississippi River mud which has been heat treated at high temperatures to resist decomposition.

Second call, 3 miles away, was at a high-pressure farm co-op and feed mill with a quickly spotted display of chemical fertilizers and sprays, and a really impressive display of farm power tools of every kind and size.

But when we said, "Organic fertilizers, bone meal, blood meal, cottonseed meal," the man behind the counter asked briskly, "How much do you want?" He also had cow manure, sheep manure, raw limestone, and soybean meal. When asked about mulches, he offered to sell us Canadian peat moss, cocoa bean shells, sphagnum moss, and shredded pine bark.

Next call, at the garden supply center of one of our largest national retail distributors, was something of a disappointment, because this is a seasonal-minded chain store operation.

On the way to the following call, the flower shop and greenhouse, we passed the city dump, the sewage plant, the brewery, and the lumberyard without stopping. The woman at the flower shop apologized "for our poor supply" but had sphagnum moss, peat moss, and cocoa bean shells available. Like the man at the big chain store, she urged us to "come back in the season; we'll have everything you can use."

The last stop of the day was to see Walter Hofsteller. Walter makes a specialty of dealing in organic fertilizers and mulches, and just about everything you need to run an organic garden and homestead.

We noted complete organic fertilizers which, although processed regionally in Georgia and Pennsylvania, are distributed nationally. There was also Sea Born, a Norwegian seaweed meal product, packaged in 100-pound sacks.

It may be helpful to list here some of the products he stocks in eastern Pennsylvania:

blood meal
feather meal
bone meal
cottonseed meal
ground granite
colloidal phosphate
rock phosphate
dolomitic limestone

The organic gardener has opened up a new market, and that market is being supplied, catered to, and courted increasingly.

Index